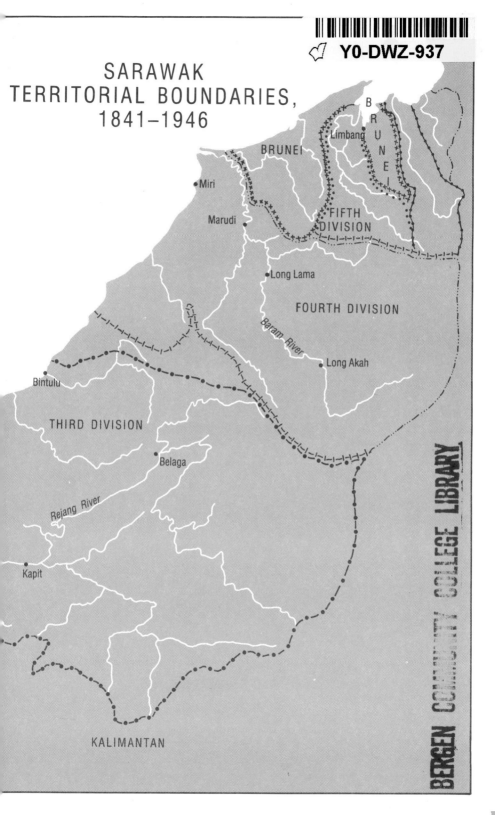

SOUTH-EAST ASIAN HISTORICAL MONOGRAPHS

Chinese Pioneers on the Sarawak Frontier
1841–1941

Chinese Pioneers on the Sarawak Frontier 1841–1941

Daniel Chew

SINGAPORE
OXFORD UNIVERSITY PRESS
OXFORD NEW YORK
1990

Oxford University Press

Oxford New York Toronto
Delhi Bombay Calcutta Madras Karachi
Petaling Jaya Singapore Hong Kong Tokyo
Nairobi Dar es Salaam Cape Town
Melbourne Auckland
and associated companies in
Berlin Ibadan

Oxford is a trade mark of Oxford University Press

© *Oxford University Press Pte. Ltd. 1990*

Published in the United States by
Oxford University Press, Inc., New York

All rights reserved. No part of this publication may be reproduced, stored in a retrieval system, or transmitted, in any form or by any means, electronic, mechanical, photocopying, recording or otherwise, without the prior permission of Oxford University Press

ISBN 0 19 588915 0

British Library Cataloguing in Publication Data
Chew, Daniel, date
Chinese pioneers on the Sarawak frontier, 1841–1941.—
(South-East Asian historical monographs)
1. Malaysia. Sarawak, History
I. Title II. Series
959.5'4
ISBN 0-19-588915-0

Library of Congress Cataloging-in-Publication Data
Chew, Daniel, date.
Chinese pioneers on the Sarawak frontier, 1841–1941 / Daniel Chew.
p. cm.—(South-East Asian historical monographs)
Bibliography: p.
Includes index.
ISBN 0-19-588915-0:
1. Chinese—Malaysia—Sarawak—History. 2. Pioneers—Malaysia—Sarawak—History. 3. Sarawak—History. I. Title. II. Series.
DS597.367.C55C43 1990
959.5'4004951—dc20
89-9403
CIP

Printed in Malaysia by Peter Chong Printers Sdn. Bhd.
Published by Oxford University Press Pte. Ltd.,
Unit 221, Ubi Avenue 4, Singapore 1440

To the memory of my father, and to my mother

Preface

THERE are compelling historiographical reasons for writing this book. There is a legacy of the stereotyped romantic view of the White Rajahs ruling a tropical island.[1] Historians, writers, travellers, missionaries, and Brooke officials have been interested in the idiosyncrasies of Brooke rule, the subject of fascination being how a European dynastic family single-handedly tamed and ruled a country inhabited by Ibans, the best-known of Sarawak's natives. The historical study of other ethnic groups in the plural society of Sarawak has been largely ignored, as if their past was of little or no consequence. One such group is the Chinese, today as numerically significant as the Ibans, and politically and economically more crucial.

It is only in recent times that studies of the ethnic Chinese have been made. Social anthropologist Tien Ju-Kang undertook a classic pioneering study of the Chinese in the state's present-day Kuching and Samarahan Divisions in 1948 and 1949.[2] What was to follow this field of Chinese studies, either wholly or substantially covering the community, were unpublished dissertations written in Western universities. Craig Lockard wrote an MA thesis, 'Chinese Immigration in Sarawak, 1868–1917',[3] in 1967, followed by a Ph.D. dissertation, 'The Southeast Asian Town in Historical Perspective',[4] in 1973. Richard Fidler conducted anthropological field-work on the bazaar Chinese of Kanowit in 1970 and 1971.[5] In 1971, a geographical study of the Sibu Foochows was carried out by Diu Mee Kuok.[6] The Sibu Foochow pioneers were also part of a wider study by Vinson Sutlive in his anthropological researches on the Ibans in Sibu.[7] The rural Chinese of the Kuching and Samarahan Divisions of the state were included by Gale Dixon in a thesis on the cultural geography of the area.[8] John Chin in 1981 wrote a short book on a general history of the Chinese in Sarawak.[9]

From a historical angle, among these published and unpublished works, Craig Lockard's contributions stand out in his historiographical and sympathetic interpretation of the experiences of the Chinese in the state. Lockard delved into the mechanics of Chinese immigration and, in his examination of the social history of Kuching, included comprehensive treatment of the Chinese in the state capital.

The historical experiences of the Chinese away from Kuching have yet to be written, and this is what this book has set out to do. A significant percentage of Sarawak's Chinese population is to be found in the rural areas, away from the main towns and urban centres.[10] I will deal with the social history of the Chinese in Sarawak at a broad level,[11] and this will comprise four categories of pioneering experiences—mining, trading, planting, and wage-labouring. An understanding of the broad historical experiences of the Chinese, especially those who pioneered and lived in the rural areas, is imperative before generalizations can be made about their role in and contributions to Sarawak society and history. Such knowledge of the Chinese in Sarawak will help to throw light on two principal themes in 'Overseas Chinese' history and historiography—Chinese–indigenous relations, and the relationship of the Chinese with the colonial regimes.

In this thematic approach of covering four broad categories of pioneering experiences, a geographical division will become evident: mining in the Bau district; trading in the Lupar, Rejang, and Baram Rivers; planting in the Lower Rejang River, spreading outwards from Sibu; and wage-labouring in the Sadong coal mines and the Miri oilfields. Archival records, supplemented by oral recollections, are used to reconstruct the pioneering history of the early settlers: mining gold in water sluices and running streams; trading in riverfront *attap* shophouses and in *kajang*-covered sampans plying the rivers; planting in small, 3- to 4-acre rubber gardens surrounded by tropical jungle; and labouring in the underground, waterlogged Sadong coal mines, and on dangerously ignitive oil-wells in Miri fronting the South China Sea. On two other levels, the book will delve into ethnic relations between the Chinese pioneers and the indigenous people, and also into the various responses of the Chinese in different combinations of time and space to White Rajah rule.

The manifest presence of the ethnic Chinese in different parts of South-East Asia has sparked off a great deal of scholarly interest in them, especially since the end of the Pacific war, during the era of decolonization and post-war indigenous nationalism. There are works which treat South-East Asia's Chinese collectively,[12] while there are others which approach them on a regional basis.[13] Books concentrating on the Chinese living in urban[14] and rural[15] areas have appeared, as have studies with geographical[16] and economic[17] themes. There are also the more specialized studies which deal with certain aspects of Chinese social organization and culture.[18] In a comparative context, I will relate my book to other studies, focusing on these pioneering experiences, Chinese–indigenous relations, and the relationship between the Chinese and colonial regimes.

Oral History Department DANIEL CHEW
Singapore
January 1989

1. There is a wide range of books and unpublished works that deal with the Brookes. Notable examples are: S. Baring-Gould and C. A. Bampfylde, *A History of Sarawak under Its Two White Rajahs, 1839–1908* (London: Sotheran, 1909; reprinted Singapore: Oxford University Press, 1989); Steven Runciman, *The White Rajahs: A History of Sarawak from 1841 to 1946* (Cambridge: Cambridge University Press, 1960); Robert Payne, *The White Rajahs* (London: Robert Hale, 1960); Joan Rawlins, *Sarawak, 1839–1963* (London: Macmillan, 1965); Conrad Cotter, 'Some Aspects of the Administrative Development of Sarawak' (MPA thesis, Cornell University, 1955); Otto C. Doering III, 'The Institutionalization of Personal Rule in Sarawak' (M.Sc. Econ. thesis, London School of Economics, 1965); Nicholas Tarling, *Britain, the Brookes and Brunei* (Kuala Lumpur: Oxford University Press, 1971); Colin Crisswell, *Rajah Charles Brooke, Monarch of All He Surveyed* (Kuala Lumpur: Oxford University Press, 1978); and R. H. W. Reece, *The Name of Brooke: The End of White Rajah Rule in Sarawak* (Kuala Lumpur: Oxford University Press, 1982).

2. Tien was a pioneer in utilizing social science approaches in studying the Chinese in Sarawak. He produced an admirable piece of work, examining the patterns of Chinese social organization structured around the role of voluntary associations and the patterns of economic relations between town (Kuching) and rural hinterland. Tien Ju-Kang, *The Chinese of Sarawak: A Study of Social Structure* (London: London School of Economics and Political Science, 1953).

3. Craig Lockard, 'Chinese Immigration and Society in Sarawak, 1868–1917' (MA thesis, University of Hawaii, 1967).

4. Craig Lockard, 'The Southeast Asian Town in Historical Perspective: A Social History of Kuching, Malaysia, 1820–1970' (Ph.D. thesis, University of Wisconsin, 1973), subsequently published as Craig Lockard, *From Kampung to City: A Social History of Kuching, Malaysia, 1820–1970*, Monographs in International Studies, Southeast Asia Series, No. 75 (Athens, Ohio: Center for International Studies, Ohio University, 1987). Lockard and Robert Pringle, who wrote *Rajahs and Rebels* (London: Macmillan, 1970), were among the first historians to write historiographically on Sarawak from 'Asiancentric' standpoints. Pringle's *Rajahs and Rebels* is a splendid work that analyses the complex relationship between the Ibans and the Brookes. Lockard, in *From Kampung to City*, traces—among other themes—the evolution of the Chinese community in Kuching, the social organizational structure of the different speech groups, the relationship between the Chinese mercantile élite and the Brookes, and the relationship between the Chinese and other ethnic groups.

5. Richard Fidler, 'Kanowit: An Overseas Chinese Community in Borneo' (Ph.D. thesis, University of Pennsylvania, 1973).

6. Diu Mee Kuok, 'The Diffusion of Foochow Settlement in the Sibu–Binatang Area, Central Sarawak, 1901–1970' (MA thesis, University of Hawaii, 1972).

7. Vinson Sutlive, 'From Longhouse to Pasar: Urbanization in Sarawak, East Malaysia' (Ph.D. thesis, University of Pittsburgh, 1972).

8. Gale Dixon, 'Rural Settlement in Sarawak' (Ph.D. thesis, University of Oregon, 1973).

9. John Chin, *The Sarawak Chinese* (Kuala Lumpur: Oxford University Press, 1981).

10. A 1980 table, below, shows the number of rural and urban Chinese together with those of other ethnic groups. Almost twice as many Chinese—61.56 per cent—live in the rural areas as compared to those in the urban centres. A high percentage of the indigenous groups also live in the rural parts of Sarawak.

Sarawak: Urban and Rural Population Distribution by Ethnic Group

	Chinese	Malays	Ibans	Bidayuh
Urban	138,580	44,209	17,824	5,129
Rural	221,973	204,548	350,684	99,785
Total	360,553	248,757	368,508	104,914

	Melanau	Other Indigenous	Indians	Others
Urban	9,746	3,592	1,365	2,084
Rural	60,067	63,560	1,929	10,478
Total	69,813	67,152	3,294	12,562

Source: 'State Population Report, Sarawak', in Department of Statistics, *Population and Housing Census of Malaysia* (Kuala Lumpur: Government Press, 1983), p. 66.

11. It is not possible to deal comprehensively with all districts in Sarawak. For instance, Lundu and many areas of the Kuching and Samarahan Divisions of the state have been ignored; likewise districts like Limbang, Lawas, and Bintulu. I believe the examples chosen for the study are representative of the major economic and social themes pursued in the book. Minor Chinese sub-ethnic groups have been left out of the study as well.

12. See Victor Purcell, *The Chinese in Southeast Asia* (London: Oxford University Press, 1965). Examples of other works are: Wang Gungwu, *A Short History of the Nanyang Chinese* (Singapore: Donald Moore, 1959) and *The Chinese Minority in Southeast Asia* (Singapore: Chopmen, 1978); William Skinner, *Chinese Society in Thailand, An Analytical History* (Ithaca, New York: Cornell University Press, 1957); and Mary F. Somers Heidhues, *Southeast Asia's Chinese Minorities* (Melbourne: Longman, 1974).

13. Examples are Skinner, *Chinese Society in Thailand*; and Edgar Wickberg, *The Chinese in Philippine Life, 1850–1898* (New Haven: Yale University Press, 1965).

14. Donald E. Willmott, *The Chinese of Semarang: A Changing Minority Community in Indonesia* (Ithaca, New York: Cornell University Press, 1960); and Tan Giok Lan, *The Chinese of Sukabumi: A Study of Social and Cultural Accommodation*, Modern Indonesia Project (Ithaca, New York: Cornell University Press, 1963).

15. William H. Newell, *Treacherous River: A Study of Rural Chinese in North Malaya* (Kuala Lumpur: University of Malaya Press, 1962); and Judith Strauch, *Chinese*

Village Politics in the Malaysian State (Cambridge, Massachusetts: Harvard University Press, 1981).

16. James Jackson, *Planters and Speculators: Chinese and European Agricultural Enterprise in Malaya, 1786–1921* (Kuala Lumpur: University of Malaya Press, 1968); and James D. Clarkson, *The Cultural Ecology of a Chinese Village: Cameron Highlands, Malaysia*, University of Chicago, Department of Geography, Research Paper 114 (Chicago, Illinois: University of Chicago Press, 1968).

17. Wu Chun-Hsi, *Dollars, Dependents and Dogma: Overseas Chinese Remittances to Communist China* (Stanford: Hoover Institute, 1967).

18. Maurice Freedman, *Chinese Family and Marriage in Singapore* (London: HMSO, 1957).

Acknowledgements

THIS study was made possible by a Murdoch University research scholarship. I am grateful to the University for its financial support and for sponsoring my field research in Sarawak. The field-work was undertaken in Sarawak between October 1980 and September 1981. This book is a revised version of a Ph.D. thesis submitted to Murdoch University in 1983.

I am indebted to a number of people who have assisted me with this study. First and foremost, I wish to express my immense gratitude to my supervisor at Murdoch University, Associate Professor Jim Warren. He has been the main driving force behind this work, from its embryonic stages right up to its completion, and finally encouraging me to have it published. Associate Professor Warren provided me with much encouragement and inspiration, as well as ideas and advice, at all stages of the study. From him I learned much of the craft of historical research and writing, for which I am grateful. I am also thankful to Dr Tim Wright of Murdoch University for his valuable and useful comments on a draft of this study. I would also like to thank Mrs Carol Warren of Murdoch University for reading and commenting upon sections of this work.

I would like to express my gratitude to the staff of Murdoch University Library, especially to Sumi Chen and Debbie Hamblin, for their kind and efficient assistance in acquiring research material for me. The assistance and friendship of Dr Shinzo Hayase at Murdoch University is also much appreciated.

In Sarawak, I am indebted to Mr Lucas Chin and Mr Loh Chee Yin of the Sarawak Museum for their encouragement and for allowing me access to the Archives. I am grateful to my family members for their assistance and encouragement. My late father gave me invaluable insights into the study and helped me with the translation of some of the Chinese sources. My brother, Alan, introduced me to many useful contacts in Simanggang (Sri Aman) and Sibu, for which I am grateful. Another brother, Kenneth, helped me with my field-work in Bau. My sister, Bernice, assisted in many ways.

I wish to thank Mr Teo Tien Teck for introducing me to interviewees and contacts in Kapit and Belaga. The generous assistance of Mr Tan Tsak Yu, Mr and Mrs Ting Tung Ming, Mr Lau Tzy

Cheng, Mr Liew Peck Kwee, Mr Robin Chung, Mr Kho Sze Kwang, Mr and Mrs Jeffrey Toh, Mr and Mrs Chai Loong Seng, Mr and Mrs John Bujang, Mr and Mrs Kung Chiu Ming, Mr Chew Kian Kiong, and Mr Chew Kian Syn, in one form or another in different parts of Sarawak, is gratefully acknowledged. I would also like to thank Mr Bong Kee Lee in Kuching, and Ms Judy Wong in Perth, for their assistance in the translation of Chinese sources. Last, but not least, I wish to express my thanks to all my interviewees in Sarawak, for this book is, in part, about them and their lives.

A Note on Spelling and Terminology

NON-ENGLISH words have been spelt the way they appear in the documents. The term 'Division' has been used in conformity with the practice in the written sources. In March 1873, Charles Brooke divided the state into three Divisions—the First Division to extend from Tanjong Datu to the Sadong River, the Second Division from the Sadong River to the Rejang River, and the Third Division from the Rejang River to Tanjong Kedurong.[1]

The Fourth and Fifth Divisions were created in 1885 and 1912 respectively from later Brunei cessions. In 1973, two additional Divisions were created out of the Third Division—the Sixth Division to cover the Sarikei area, and the Seventh Division to include the Kapit district. All the Divisions were again renamed in 1987 as the Kuching, Samarahan, Sri Aman, Sarikei, Sibu, Kapit, Bintulu, Miri, and Limbang Divisions.

To avoid confusion in the terminology concerning Divisions, I have used river basins to describe the appropriate areas. The 'Batang Lupar basin' refers to the Brooke usage of the term, 'Second Division'. Similarly, 'the Rejang' means the Third Division, 'the Baram' refers to the Fourth Division, and the First Division includes the present-day Kuching and Samarahan Divisions. Where the terms 'First Division', 'Second Division', 'Third Division', and 'Fourth Division' are used in the book, they conform to the Brooke usage of these terms. Old place-names used in the sources and used previously in common parlance have also been retained. 'Simanggang' is used instead of 'Sri Aman'.

---

[1] *Sarawak Gazette* (*SG*), 1 March 1873, p. 121.

Contents

Preface	vii
Acknowledgements	xii
A Note on Spelling and Terminology	xiv
Appendices	xvii
Figures	xvii
Maps	xviii
Tables	xviii
Plates	xix
Abbreviations	xxi
Weights, Measures, and Currencies	xxii
Introduction	1
A Note on Sources and Methodology	1
The Physical Landscape	3
Ethnographic Background	6
1 Mining Pioneers in Bau, 1800–1857	18
Kongsi Traditions in Kalimantan	18
The Origins and Growth of the 'Twelve Company' in the Bau Mining District	23
Early Political, Economic, and Social Conditions of the 'Twelve Company' in Bau	25
Kongsi–Brooke Rivalry	29
2 Economic Changes in Bau, 1857–1898	37
The *Kongsi*–Brooke War of February 1857	37
Revitalized *Kongsi* Mining	38
The Brooke Presence in Bau	41
The Hakkas and the Borneo Company	42
The Shift to Crop Cultivation	43
3 Traditional Patterns of Trade in Borneo	50
The Historical Patterns of Chinese Trade with Borneo	50
Traditions of Inter-island, Coastal, and Riverine Trade in Borneo	55

4 **Chinese Traders in the Lupar, Rejang, and Baram River Basins: The Origins and Pioneers of *Ulu* Trade** 63
  The Origins and Expansion of Trade 63
  Pioneering Life 77

5 **Organization of *Ulu* Trade: Economic Exchange and Transformation** 100
  Products, Procurement, and Exchange 100
  The Transformation of the Riverine Economies 108
  The System of Credit 114

6 **Up-river Trading and Social Relations** 122
  Trading Regulations and Restrictions on the Chinese 122
  Ethnic Relations and Riverine Trade 130

7 **The New Foochow Colonists: The Beginning of the Rubber Cash-crop Economy in the Lower Rejang, 1901–1920** 142
  Wong Nai Siong and the Sponsorship of Foochow Migration 142
  Pioneering Conditions in the Lower Rejang 145
  Social Organization 149
  The Rubber Economy 152

8 **The New Foochow Colonists: Social Relations and Economic Transformation, 1921–1941** 160
  The Spread and Expansion of Foochow Settlements 160
  The Land Disputes 161
  Economic Background 170
  Social Organization 174

9 **Chinese Labourers and the State Economy, 1870–1941** 181
  The Opening of the Sadong Coal Mines 181
  The Significance of the Sadong Colliery to the Brooke State 191
  The Opening of the Oilfields in Miri 192

10 **The Chinese and the Brookes** 203
  Indirect Brooke Rule over the Chinese 203
  The Chinese and the Legal Structure 207
  Brooke Reliance upon the Chinese as a Source of Revenue 211

Conclusion 219

Glossary	232
Appendices	236
Bibliography	258
Index	274

Appendices

1	List of Interviewees	236
2	Trade on the Kalaka and Saribas Rivers, 1907	244
3	Boat Cargo, 1922	246
4	The Burning of Sibu Bazaar, 1889 and 1928	247
5	Rejang Settlement Order, 1880	249
6	Agreement between Charles Brooke and Wong Nai Siong for the Immigration of Foochows, 1900	250
7	Cultivation of Rice in the Rejang	253
8	Opening of the New Chinese Court, 1911	256

Figures

1	The Domestic Impact of *Ulu* Trade	101
2	The Regional Economy, *Ulu* Trade, and the Flow of Goods	107

Maps

Sarawak Territorial Boundaries, 1841–1946	*Front endpapers*
Present-day Administrative Divisions of Sarawak	*Back endpapers*
1 The Provinces of Kwangtung and Fukien Showing Approximate Places of Origin of Different Dialect Groups	7
2 *Kongsi* in West Borneo and Sarawak	19
3 Bau Mining District	22
4 Archaeological Sites in Sarawak	51
5 The Lupar River Basin	64
6 The Rejang River Basin	70
7 The Baram River Basin	73
8 The 'New Foochow' District	143

Tables

7.1 Foochow Arrivals, Land Titles, and Rubber Prices, 1901–1920	154
8.1 Foochow Arrivals, Land Titles, and Rubber Prices, 1921–1940	171
9.1 Sadong Coal Account, 1881–1899	191
9.2 The Miri Oilfields' Labour Force, 1919 and 1920	194
10.1 Farms Revenue for Selected Years between 1873 and 1896	213
10.2 Farms and Customs Revenue, 1925–1939	214
10.3 Import and Export Duties, 1876–1886	215
10.4 Comparisons between Customs and Farms Revenues, 1900–1903	216
10.5 Import and Export Duties, 1910–1920	216

Plates

Between pages 202 and 203

1. *Kongsi*-house of the 'Fifteen Company' in Marup, near Engkilili.
2. Interior of the *kongsi*-house of the 'Fifteen Company'.
3. Flag-poles symbolizing the autonomy of the 'Twelve Company' in Bau. According to oral tradition, the *kongsi* flag was raised here every morning. *Kongsi* members sentenced to death were executed here in front of the flag.
4. A late nineteenth-century photograph of Chinese miners in a gold mine. (Courtesy Sarawak Museum.)
5. A temple built in honour of Liew Shan Pang, leader of the 'Twelve Company' in Bau. This temple was rebuilt in 1976. It marks the spot where Liew was supposed to have been buried.
6. Chinese traders buying camphor from Kayans. (From W. H. Furness, *The Home Life of Borneo Headhunters, Its Festivals and Folklore*, Philadelphia: Lippincott, 1902.)
7. Ibans bargaining over the sale of valuable Chinese jars. (From W. H. Furness, *The Home Life of Borneo Headhunters, Its Festivals and Folklore*, Philadelphia: Lippincott, 1902.)
8. Ibans on a trip to a bazaar. Contact with Chinese traders has brought about changes in material culture for the Ibans. The man on the right is wearing a fez and a Chinese shirt. The woman who is standing is wearing a skirt made of black *blachu*. (From R. H. W. Reece, *The Name of Brooke: The End of White Rajah Rule in Sarawak*, Kuala Lumpur: Oxford University Press, 1981.)
9. Iban clients together with Chinese traders in a longhouse. (Courtesy Sarawak Museum).
10. A Chinese boat hawker in front of his boat, 1981.
11. A late nineteenth-century view of the Kuching waterfront. (Courtesy Sarawak Museum.)
12. A longhouse. (Courtesy Sarawak Museum.)
13. A Chinese trader in Belaga, 1981.
14. A rattan raft being floated down to a bazaar. (From W. H. Furness, *The Home Life of Borneo Headhunters, Its Festivals and Folklore*, Philadelphia: Lippincott, 1902.)

15 Marudi bazaar, 1902. (From W. H. Furness, *The Home Life of Borneo Headhunters, Its Festivals and Folklore*, Philadelphia: Lippincott, 1902.)
16 Chinese ceramics. (Courtesy Sarawak Museum.)
17 Sing Chio Ang (Sungei Merah), Sibu, 1903. Sungei Merah was one of the two sites for the first batch of Foochow settlers. (Courtesy Ling Kai Cheng.)
18 The second home of Reverend and Mrs Hoover in Island Road, Sibu, 1905–8. (Courtesy Ling Kai Cheng.)
19 The first Anglo-Chinese School in Sibu, 1903. (Courtesy Ling Kai Cheng.)
20 The first quarterly Methodist Conference, 1903. (Courtesy Ling Kai Cheng.)
21 The first Methodist chapel in Sibu at Sing Chio Ang (Sungei Merah), 1903. (Courtesy Ling Kai Cheng.)
22 A rubber smallholding. (Courtesy Sarawak Museum.)

Abbreviations

BB	*Borneo Bulletin*
BMJ	*Brunei Museum Journal*
BPB	Baram Probate Book
CCB	Court Case Book
CO	Colonial Office
CSSH	*Comparative Studies of Society and History*
FO	Foreign Office
HMSO	Her Majesty's Stationery Office
JAS	*Journal of Asian Studies*
JIAEA	*Journal of the Indian Archipelago and East Asia*
JMBRAS	*Journal of the Malayan/Malaysian Branch, Royal Asiatic Society*
JSBRAS	*Journal of the Straits Branch, Royal Asiatic Society*
JSEAS	*Journal of Southeast Asian Studies*
MSS Pac	Manuscripts kept at Rhodes House, Oxford
SA	Sarawak Archives
SG	*Sarawak Gazette*
SGG	*Sarawak Government Gazette*
SMJ	*Sarawak Museum Journal*

Weights, Measures, and Currencies

Weights

1 *kati* ≡ 1⅓ lbs
1 *pikul* ≡ 100 *kati* ≡ 133⅓ lbs
1 *koyan* ≡ 40 *pikul* ≡ 5,333.3 lbs

Measures

1 *bunkal* (gold measure) ≡ 1 gold *tahil*
1 *gantang* (dry weight) ≡ 8 *kati*
1 *passu* ≡ 8 *gantang*
1 *depas* ≡ arm's length
1 *gelong* ≡ coil of rattan in bundles, approximately 100 vines

Currencies

The Sarawak dollar was tied to the Straits Settlements dollar and tended to fluctuate in value

Introduction

A Note on Sources and Methodology

THE field-work for this research was undertaken over a period of eleven months from October 1980 to September 1981. Archival research was first carried out at the Sarawak Museum Archives in Kuching, Sarawak. This was then followed by extensive travelling in many parts of the state to interview elderly Chinese pioneers and their descendants.

The significant unpublished sources consulted for this study include the Second Rajah's Letter Books, Agreement Books, Order Books, Court Records, District Records, and Probate Books. The Letter Books contain the correspondence between the Rajahs and their outstation European officers which from time to time directly concerned the Chinese in different parts of Sarawak. The Agreement and Order Books are an important source of information on administrative matters, containing agreements, rules, regulations, and orders, most of them of an *ad hoc* nature. The Court Records hold invaluable information on legal and administrative matters which give some insight into the economic and social lives of the *ulu* (rural) Chinese, as well as indicating Brooke policies towards them. In Sibu, the Land Office files were consulted. The land title deeds contain useful information on Foochow settlement patterns.

A published source, the *Sarawak Gazette* (*SG*), started in 1870, has been used extensively for this book. It was a government publication, compiled monthly and, at times, at fortnightly intervals. The paper carried mostly official news and articles likely to interest the European community. Nevertheless, certain Brooke officials—namely the Residents and District Officers—who were required to submit the reports made observations on local happenings and the local population. Such reports, which provided an outsider's view of the Chinese, have to be read critically. The *Sarawak Government Gazette* (*SGG*), first published in 1908, contains information on government announcements, district annual reports, and statistics on revenue and expenditure.

The Archives hold a fine collection of Bornean manuscripts and books, such as the published correspondences of James Brooke, travelogues, missionary memoirs, and the reminiscences of Brooke officials. Though most of these books concentrate on Brooke rule and its paternal relationship with the Ibans, they make occasional references to the Chinese.

Chinese publications by various speech-group associations and by amateur historians such as Lau Tzy Cheng and Liew Peck Kwee have been helpful. There is, however, a gap in Chinese sources. Chinese newspaper publishing in Kuching and Sibu started late in the 1920s and 1930s[1] and, although most of the papers have disappeared, the Sarawak Museum has microfilmed some copies of the pre-war *Sibu Chinese Daily News* and the *Overseas Chinese Daily News*.

A crucial component of my research methodology was oral history interviewing, collecting reminiscences and stories handed down from the past, from elderly interviewees. There are gaps and limitations in the written sources. The archival material created by the Brookes and their officers had an inherent bias, besides giving insufficient information on the Chinese. This is where the oral sources are valuable in complementing the written material and presenting a perspective from within.

I have used some oral accounts collected by Lau Tzy Cheng and Liew Peck Kwee. It was necessary for me to travel widely in Sarawak to record the memories of the elderly Chinese pioneers or their descendants. It was a time-consuming and physically demanding experience requiring long, slow journeys by bus or boat, or on foot. Oral history interviewing with persons belonging to different speech groups requires considerable linguistic skills. I personally conducted Hokkien and Teochiu interviews. With interviewees who were Hakka or Foochow, it was sometimes possible to use Hokkien or Teochiu, but, more often than not, I had to rely on interpreters.

A major difficulty faced by reseachers in collecting oral history information is to gain the confidence of potential informants. I found it useful if I was given the names of potential contacts in advance or if I was introduced to leading persons residing in particular places. To put potential interviewees at ease, I had to explain clearly my objectives and use a mixture of tact, patience, and persistent prompting.

I first started my interviews by using a questionnaire and a tape-recorder. It did not take me long to discover that a tape-recorder can indeed be an obtrusive intruder for people in remote rural communities and can raise suspicion in their minds. Soon, I discontinued using the tape-recorder. Formal questions which I asked in a set pattern produced abrupt, terse answers, and I found these unhelpful. I then resorted to informal, open-ended interviews and my role was to prompt the interviewees with topical questions, otherwise irrelevant trivia would have cropped up. By this method, informants felt more comfortable and more useful information was obtained than could otherwise have been gleaned from a formal question-and-answer session.

A recurring theme which ran through the interviews was that of pioneering hardship, which most of the interviewees tried to impress upon me. Sarawak is a physically harsh country and my observations of the ethnographic present, together with the conditions under which I conducted my interviews, gave me an inkling of previous circumstances and enabled me to empathize with my interviewees. Sometimes, interviews had to be carried out with my feet stuck ankle-deep in the mud along a river bank, in coffee-shops[2] with curious onlookers standing near by, or by the dim light of a kerosene lamp at night. My extensive travelling throughout Sarawak provided me with important insights into my research, allowing me to appreciate a jungle landscape traversed by rivers. I also visited rubber gardens near Sibu, old gold mines in Bau, an abandoned *kongsi*-house near Simanggang, and deserted oil-well sites in Miri, to get a feel of what the past could have been like for Sarawak's Chinese pioneers.

I was a participant observer. By seeing at first hand present living conditions, circumstances, and work, I was able to visualize the continuity from the past. I slept on the top floor of shophouses in the way Iban clients in the past used to do, when they came down to the bazaars (and even now continue to do). Rural dwellings were visited, which have changed remarkably little over time, with mud floors and *attap* roofs. There were many opportunities to sample local hospitality, and many informants invited me to share their meals. I ate local delicacies like raw fish, sago pellets, turtle, wild boar and deer meat, jungle ferns and fruits, and various types of river fish, cuisine seldom to be found in the big towns. On one occasion, on a boat trip, the boat-owner asked me to join him for a simple meal of canned sardine and glutinous rice cooked at the back of the boat over a kerosene stove. The extensive travelling and the field-work helped to reinforce my archival research in the Sarawak Museum, giving me a better 'feel' for the documents.

The Physical Landscape

An understanding of the physical backdrop of the vast island of Borneo is necessary in order to appreciate how the Chinese moved into the empty frontiers of Sarawak and adjusted themselves to their new environment. Borneo is the third largest island in the world and Sarawak, with its present political boundaries, has an area of 48,250 square miles.

The political divisions of Sarawak, created largely by Rajah Charles Brooke, conform to a natural watershed separating the state from the rest of Borneo. The island was renowned for its mountains, jungles, swamps, and rivers—unattractive geographical conditions at first sight—but these were not a deterrent to the pioneering Chinese. Some pockets of Sarawak were mineral-rich, and there were other areas suitable for the planting of rubber. However, in most of Sarawak, an imposing jungle and a multitude of rivers presented neither mining nor planting opportunities, and a different kind of livelihood in trading had to be taken up instead.

Sarawak is geographically divided into an alluvial coastal plain, a rising mountainous interior, and an intermediate belt of undulating country. The mountains vary from 2,500 feet to 4,000 feet in height. The intermediate country is deeply dissected hilly land. Peat and mangrove swamps cover the coastal plain and river valleys reaching out to the sea.[3]

Most of central and northern Sarawak between the coast and an elevation of 1,500 feet has a thick forest canopy which occurs over very thin soil and is sustained by a hot and moist climate. The forest is dense and an overhanging cover shuts out the sunlight. In this dense vegetation, especially in the headwaters of the Rejang and Baram Rivers, thrive wild rubber, camphor trees, beeswax, and *engkabang*, all valuable products in the nineteenth century.

Torrential rainfall, averaging 120–160 inches per year, is responsible for heavy soil erosion in the western part of Sarawak, in Lundu, in higher parts of the belt of hilly country, and in the Upper Rejang and Baram Rivers. The soil cover is generally thin and sandy, unsuitable for agriculture, and native swidden cultivators can plant only one or two crops of padi before being forced to move elsewhere. The soil in the Lower Rejang around Sibu is of this type—thin, sandy, and infertile, and yet it was this area which Charles Brooke hoped Foochow migrants would convert into the 'rice bowl' of Sarawak.

Mangrove swamps cover the deltas of the Sarawak, Rejang, and Trusan Rivers. Peat swamp is widespread between the Sadong and Saribas Rivers, from Kuala Paloh to Bintulu, and in much of the lower Baram valley. These swamp-lands naturally discourage human habitation, although Malay fishing villages can be found. However, in the past, the coastal sago palm which thrives well in the silted and swampy conditions of central Sarawak, along the Mukah coast, has been important as a food staple and as an industrial starch for export.

The immediate coast and the interior were unsuitable for pioneering purposes. More attractive places were to be found inland from the inhospitable river deltas. There is access to the sea and the up-river hinterland. It was at these places that major towns of the nineteenth century developed—towns such as Kuching, Simanggang (Sri Aman), Marudi, and Simunjan. Brooke policies on Chinese migration encouraged the growth of Sibu, while Miri was a by-product of the oil industry. These towns all share a common character, originating as mixed Malay or Muslim–animist native settlements and, over time, being transformed into Chinese-dominant centres. While the Chinese have helped to build and populate the towns, they have also moved into the interior to open up new frontiers through their mining, trading, planting, and labouring endeavours.

By far the most important geographical feature of Sarawak is the river. The main rivers rise in the interior mountain ranges, cascading down rocks near the source, meandering across the undulating hilly country, and slowly spilling out to the sea at the river deltas. The different personalities of these rivers define the political and social character of the administrative divisions they represent. As one goes up-river, rapids and rock obstacles are ever-present dangers. In the dry season, during the middle of the year, when the water-level drops and the rocks are exposed, few boats would risk 'shooting' across rapids like the well-known Pelagus rapids between Kapit and Belaga. On the other hand, during the end-of-the-year monsoon, river levels can rise suddenly with heavy downpours and make river travelling dangerous as currents are swift. Whether one wanted to travel up-river against the current or down-river with the current would influence the duration of a boat journey.

Distances between places are not calculated in absolute terms, but rather by the measure of so many days' journey between one spot and another; the estimate varies according to the level of the river, the season, and the direction one is going. Information, goods, and people going from one destination to another flow as quickly or as slowly as the rivers. The arrival of boats breaks the monotony of daily life in isolated communities and seldom fails to attract attention or an enthusiastic welcome. The mooring of a boat at a landing place near a bazaar or a longhouse has always been a social occasion, as people isolated from each other and from the outside world converge to help load and unload, welcome the visitors, exchange gossip, or just be silent observers of the whole scene. This may partially explain why the traders of old, when they felt it safe to do so, would venture into unknown and remote places with the knowledge that they would be welcomed. The economic and social influence exerted on Sarawak's communities by a muddy brown river, with its ebb and flow, is indeed great.

Ethnographic Background

An uninviting and inhospitable island environment has insulated most of North-west and West Borneo[4] from the outside world. The exception is the coast and its immediate hinterland which received the currents of material culture, social ideas, and religious values from China, India, and the Malay archipelago. The influence of China is evident in the material culture of the natives, in ceramics which are highly revered for social and cultural reasons, and in local artistic styles. Islam could have reached the coast of North-west Borneo by the end of the fourteenth century[5] and spread along the littoral coast to Sarawak. Ideas of kingship and statecraft could have come from the Majapahit and Srivijaya empires. The coast, which first felt the impact of external influences, in turn transmitted ideas, values, and material artefacts up-river. This was a slow process as most of the interior remained cut off from the outside world.

One of the more significant local developments was a dichotomous Muslim–animist relationship which coincided with the geopolitical character of down-river ports exerting control over up-river tribal communities.[6] Where considerable demands concerning trade, tribute, and tolls were imposed on the up-river people, there was resistance; alternatively, the longhouse communities would relocate themselves even further up-river, so as to be out of range of the forces of change and control emanating from the coast.

Against this Muslim–animist geopolitical background, significant happenings were to take place after 1800. The Chinese arrived in the Bau district early in the nineteenth century to mine gold. They were followed shortly by James Brooke. The presence of the Chinese and Europeans upset the existing pattern of political and economic relations. In place of Muslim control on the coast was substituted European might, backed by formidable military technology and stiffened with native allies. The late nineteenth- and early twentieth-century spread of the pioneering Chinese diaspora, down-river, up-river, and along the coast, in mining, planting, trading, and labouring occupations, changed the erstwhile pattern of Muslim–animist economic and social relations.

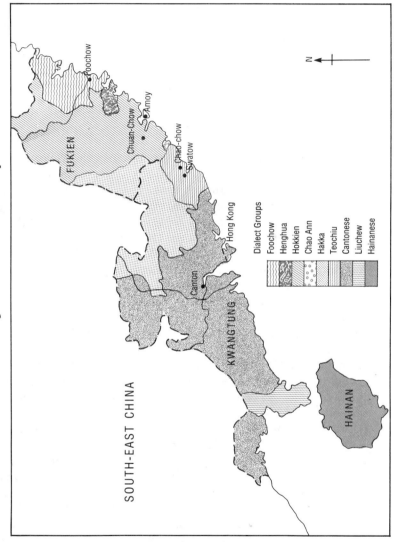

Map 1 The Provinces of Kwangtung and Fukien Showing Approximate Places of Origin of Different Dialect Groups

The Chinese

The majority of Chinese settlers in Sarawak come from the provinces of Kwangtung and Fukien on China's south-east coast. These two provinces contain heterogeneous speech groups, most of which are also found in Sarawak. Different dialects, different cultural traits, and diversity in religious beliefs and practices are to be found among the various groups in South-east China. Foochows, Hakkas, Hokkiens, Teochius, and Cantonese are predominant in Sarawak. Their diverse backgrounds are reflected in their different forms of economic and social adaptation.

Hakkas (Kheh)

They have a migratory history, originating from the Yellow River valley in northern China and out of which they started moving in the fifth century. By the thirteenth century, they had settled to the north-west of Chao-chow in Kwangtung.[7] Gradually, they came to occupy a belt of mountainous terrain from Fukien to northern Kwangtung, to Kiangsi and Hunan. The Hakkas in Kalimantan and Sarawak originate from the four Hakka districts in Kwangtung province—Chia-ying (Kei Hsien), Tapu, Kityang (Hoppo), and Huilai (Hailu-feng).[8] The four dialects spoken in these four districts are mutually intelligible. The term 'Hakka' is not derived from any territory, and the popular theory is that it means 'guest', a reference to the migratory habits of these people. Hokkiens and Teochius call Hakkas 'Kheh', by which term they are now commonly known. Over time, the Hakkas themselves have come to accept this label.

But the label carried more than just a name. Being late intruders to Kwangtung, Hakkas were regarded by the Cantonese, Teochius, and others to be of an inferior status, a prejudice accentuated by the fact that they inhabited inhospitable and infertile mountainous terrain. Other cultural characteristics which set the Hakkas apart from other speech groups were the high status of women in Hakka society and the fact that Hakkas did not bind their women's feet. The Hakkas, with their strong rural inclination, were to set up mining *kongsi* in Kalimantan and Sarawak and, later in the nineteenth century, they took up cash-cropping. The Hakkas are now concentrated in the Kuching and Samarahan Divisions, with a smaller number in the Sri Aman (Simanggang) area, and in Miri. The majority of Hakkas still live in the rural areas, being cash-croppers and market gardeners.

Foochows

Foochows originate from Foochow city and its surrounding counties in northern Fukien. The whole Foochow prefecture consists of ten *hsien* (counties): Minhsien, Liengchiang, Loyuan, Haukwong, Changlok, Fuching, Mintsing, Yungfu, Jutien, and Pingnan.[9] There are only slight differences in the dialects spoken in these counties. The Foochows are clustered along the Min River, surrounded by mountainous terrain, although Foochow city and its immediate environs are in a fertile area. Most Foochows are peasant farmers and converted Methodists, a background which was to assert itself in Sarawak.

Foochows, together with the Hakkas, form the two largest Chinese sub-ethnic groups in the state. Foochows are found mainly in the Lower Rejang, especially in the triangular region bordered by Sibu, Binatang, and Sarikei, but they have recently spread out to other areas of the state such as Kuching, Sri Aman, Marudi, Mukah, and Limbang. Among the different speech groups, they have a reputation for being hard-working and for being insular. Among the Chinese, they are perhaps the most adaptable and most economically aggressive, characteristics which have earned them the envy of others.

Hokkiens

This group is from the Amoy area in the southern part of Fukien province. The name 'Hokkien' comes from the pronunciation in that dialect of the provincial name, and Hokkiens are also known as Fukienese. The *émigré* Hokkiens are mostly from Ch'uan-chow and Changchow. Ch'uan-chow was a trading port, and when the Manchus gained power, the banning of foreign trade could have caused the Hokkiens to go abroad, either to emigrate or to trade.[10] This might explain why the Hokkiens were one of the earliest groups in Sarawak during the period of Brooke rule. Hokkiens have a predilection for urban living, which accounts for their strong presence in Kuching. Along with a preference for urban life was an inclination for entrepreneurial activities, from the big export–import shops of Kuching, to outstation trade, and to the far-flung boat hawker of the interior. Hokkiens have been economically dominant in trade and are to be found in almost every town in Sarawak. They form the third-largest Chinese sub-ethnic group in the state.

Chao-anns (Chawan)

This small group hails from Chao-ann *hsien* in Fukien province. Its dialect is intermediate between Hokkien and Teochiu but, in economic and cultural terms, they are more aligned towards the Hokkiens.[11] Their small number belies their economic importance for they have been predominant in places like Dalat, Mukah, Sibuti, and Tatau.

Teochius

Teochius reside in the Swatow region of North-east Kwangtung, and in southern Fukien. Their dialect bears some resemblance to Hokkien. In Sarawak, they form the fourth-largest Chinese dialect group. They share with Hokkiens a preference for urban life and for trade, and are to be found in Kuching, Sri Aman, Mukah, and Bintulu, principally as traders, although smaller numbers have taken up planting.

Cantonese

Cantonese-speakers are to be found in Canton, Macau, Hong Kong, and throughout Kwangtung province. Few Cantonese have come to Sarawak, and those who have, are to be found in the Rejang delta area around Sarikei, originally involved in timber logging and then in pepper planting, and in Miri where they worked as labourers in the oil industry.

Other Minor Chinese Dialect Groups

Henghuas come from Henghua and Sienya *hsien* in Fukien. Throughout Sarawak, they have been dominant in fishing and in running bicycle shops. In Sarawak, the Hainanese—or Hailams as they are more commonly known—form a small, highly urbanized group associated with coffee-shops, restaurants, and the catering business. There are also small groups of Southern Mandarin speakers—the Hupeis who are 'tooth-artists', and the Kiangsis who are expert furniture-makers.

Native Ethnic Groups

The ethnic classification of natives has always presented problems. Before the advent of the Brookes, longhouse or village communities identified themselves according to geographical locality—usually a river, part of a territory, a longhouse name—or after prominent persons, such as leaders and aristocrats. There is a difference between what the diverse groups called themselves and what others labelled them.

The use of external labels or exonyms is a fairly recent practice, but it has had a permanent impact on ethnic classification. Exonyms were used by the Brookes and their officials to give some kind of order and to confer ethnic status on the numerous groups they encountered. The natives themselves were to apply arbitrary classifications on one another.

A good example of the use of exonyms is the word 'Dayak', adapted from 'Dajak' by which the Dutch called all natives in Kalimantan. James Brooke used it to refer to non-Muslim indigenes. He applied the term 'Sea Dayak' to the marauding natives of the rivers of the Second Division.[12] A locally used exonym is 'Iban' (Ivan or 'wanderer'), employed by Kayans to refer contemptuously to their restless and volatile Dayak neighbours in the Upper Rejang.[13] The significance of the use of exonyms is that those referred to as such have, over time, accepted their usage. Nevertheless, there are communities which are called after place-names or toponyms—examples of such communities being the Kayans, named after the Kayan River in Kalimantan.

Confusing though the ethnic nomenclature is, it has been decided for the purpose of this book to use the more commonly recognized and accepted ethnic terms of 'Malay', 'Dayak' (Sea and Land), 'Iban', 'Kayan' and 'Kenyah', with the aim of avoiding further confusion. Where possible, toponyms will be used to identify the groups with particular river basins or districts.

Malays

It has been conventionally thought that the Muslims in Kuching and in the riverine deltas of the Kuching, Samarahan, and Sri Aman Divisions are Malays sharing common origins with their counterparts in Malaya and Sumatra. It is now widely accepted that the Malays in Sarawak are of diverse origins, coming from places such as Java, Celebes, and Brunei, as well as comprising local communities converted to Islam. Becoming a Muslim in Sarawak was referred to as '*masok Melayu*' (become Malay), entailing an abandonment of one's old religion and culture and the adoption of Malay cultural values. Being a Malay became synonymous with being a Muslim. But the opposite did not necessarily hold and, today, there are Melanaus on the Mukah littoral coast who are Muslims.

Pre-Brooke Malay society was stratified. At the apex were those who claimed aristocratic connections and genealogical links with Arabia. Many of the Malay leaders were active traders, who often combined their high social status with trade and political ambitions. James Brooke utilized the talents of the élite Malays as perfunctory advisers and native officers, and generally discouraged them from trading.

Ibans (Sea Dayaks)

The Ibans are the best-known of the native people in Sarawak, with their past reputation of being feared head-hunters. The Ibans are today spread all over Sarawak, partly out of their own volition and partly due to official encouragement. In the nineteenth century, certain common characteristics which distinguished the Ibans from other indigenous people were longhouse residences, the practice of swidden cultivation, animistic beliefs, and a non-stratified social structure.

There were, however, certain Iban cultural and social traits which brought them into regular exchange arrangements with Chinese traders. Through long-standing contact with the Chinese, the Ibans acquired a liking for ceramics and earthenware, using jars, plates, and bowls for utilitarian purposes such as brewing rice wine, and as containers for medicine, ointments, and jewellery. The ownership of these items, handed down as heirlooms, conferred much social prestige and was a sign of material prosperity. Even in the late twentieth century, ceramics and earthenware play an important role in the ritualistic and religious aspects of Iban life.[14] Contact with Chinese traders linked the Ibans with market forces, entrenching them within a regional economy stretching from Singapore to Kuching, the up-river bazaar, and finally to the longhouse. The Ibans developed a taste for luxury items as well as essential commodities such as iron, cloth, tobacco, and salt. A more macabre aspect of cultural contact between Chinese trader and Iban was head-hunting, and in the past many an itinerant trader fell victim.

To the Ibans, the most important of all material possessions is land. According to Iban mythology, the land belongs to the spirit Pulang Gana who received a sod of earth for his inheritance and thus became the Spirit of the Earth. The Ibans obtain spiritual permission to clear and farm the land through sacrificial offerings and observance of rituals. Similarly, they acquire general hunting and gathering rights to a particular area, but they do not own the land.[15] The Iban idea of land ownership and use differed markedly from that of the pioneering Foochows in the Lower Rejang, resulting in disputes, litigation, and open conflict.

Bidayuhs (Land Dayaks)

The ancestors of the Bidayuhs were the non-Muslim native people in the north-western corner of Sarawak, whom James Brooke first encountered. They were known as 'Land Dayaks', initially called 'hill tribes' by the White Rajah.[16] They lived on hilltops for protection against head-hunting enemies. They were quite different in temperament, cultural characteristics, and language from the more aggressive and versatile Ibans. Nevertheless, like the Ibans, they collected ceramics and earthenware for utilitarian as well as religious purposes, and for prestige. They were also dependent on Chinese traders for goods from outside. The Bidayuhs observe land *adat* rules similar to those of the Ibans.

Melanaus

The Melanaus refer to themselves as '*a-likou*', meaning 'people of the river'.[17] The Melanaus claim that their present name was given to them by the Brunei Malays.[18] Previously, they had a stratified society of aristocrats, commoners, and slaves, different from the non-élitist Ibans, but similar to the Kayans and Kenyahs of the Upper Rejang. They used to live in longhouse-type dwellings which have since disappeared. This could be due to the influence of Islam. Many Melanaus converted to Islam due to a strong Brunei Malay presence on the Mukah coast. The Brunei Malays were attracted by the lucrative trade in sago in the nineteenth century. The sago trade was also to attract Chinese traders to the same area in the latter half of the nineteenth century. Melanaus, like native people elsewhere in Sarawak, revere old ceramics. Kept as heirlooms, bowls and jars are principally used by the Melanaus as *berian* (bride-wealth), and for burial purposes.

Kayans and Kenyahs

Kayans and Kenyahs share a complexity of social, economic, and cultural characteristics which include the shifting cultivation of hill padi, longhouse residence, and stratified social systems with hereditary chiefs.[19] Presently, they are located in the valleys of the Upper Rejang and Balui, the Baram and Ulu Kemana Rivers.

Like the Ibans and the Melanaus, the Kayans and Kenyahs treasure prestige goods like jars and beads. Dragon-adorned jars used to be buried in coffins with Kayan and Kenyah aristocratic chiefs and, even today, are believed to possess magical powers: water placed in the jars is used to treat sickness and to sprinkle over rice-fields.[20] In the past, ancient glass, porcelain, and faience beads were the most coveted of all forms of material wealth, and served as a medium of exchange and as ceremonial objects. Beads were worn by the wives of chiefs on ceremonial occasions.[21] As with the Ibans, the economic forces of trade introduced the Kayans and Kenyahs to the Chinese, on whom they came to rely for their ceramics, beads, luxury needs, and foodstuffs.

Other Minority Ethnic Groups

Other remaining ethnic groups are the Kajangs, Kedayans, Kelabits, Muruts, Bisayas, and Penans.[22]

In the Kajang group of the Upper Rejang are such diverse groups as the Tanjongs, sedentary Punans (Punan Ba), Skapans, Kajamans, and Lahanans. They are longhouse-dwellers and although linguistically and culturally related to the Melanaus on the Mukah coast, they are more influenced by their immediate Kayan and Kenyah neighbours.

The Kedayans are a Muslim group with strong ties in Brunei, and they are found at intervals along the coast of the present Miri and Limbang Divisions where they cultivate padi.

Two related groups, the Muruts and the Kelabits, inhabit the interior valleys of northern Sarawak. The Muruts concentrate in the Trusan valley, while the Kelabits are to be found in the headwaters of the Baram and in the Bario highlands, some 3,000 feet up. Both communities have a reputation for being efficient padi farmers.

Bisayas are located in the middle Limbang valley. Scattered in the deep forests of the Upper Rejang and the Baram are the nomadic Penans.

Groups as diverse as all these have some kind of contact with Chinese traders, and the Kelabits, for example, in their remote Bario highlands are known to keep old Chinese ceramics.

Europeans

The final element in the plural Bornean society of Sarawak in the period under discussion is the European community, epitomized by the Brookes. The founding of the Brooke Dynasty in 1841 is a well-recorded and well-written annal in Sarawak's history.[23] It will be briefly summarized here.

The Sarawak River basin, from which the present-day state takes its name, was in its pre-Brooke days a nominal part of the Brunei Sultanate. This nondescript river basin existed quietly until the discovery of antimony ore aroused Brunei interest early in the nineteenth century. With this discovery and knowing that the Land Dayaks could be coerced to mine the ore, Pengiran Makota, a Brunei prince and nominal ruler of the district, took up residence in Sarawak sometime between 1824 and 1830. Pengiran Makota's presence and his taxation demands provoked resistance from the local Malays and the Land Dayaks. The conflict reached a stalemate, whereupon the Brunei Sultan sent his uncle and prime minister, Pengiran Muda Hashim, to Kuching. However, Pengiran Muda Hashim's presence made little difference to the situation for the fighting between the warring factions remained at a standstill when James Brooke, an English adventurer, arrived in 1839.

Brooke was an idealistic young man who had retired early from the Indian army. It was his love of adventure in the East that took him to Borneo, among other places. He was very much influenced by the Rafflesian idea of a strong British presence in the Indonesian archipelago. He had sailed to Sarawak in July 1839 after having learned that the Brunei prince was favourably disposed to the English and that the district had valuable antimony ore. Brooke intervened in the civil war by helping Pengiran Muda Hashim, in return for which he demanded and obtained the governorship of Sarawak, roughly the area of the present-day Kuching and Samarahan Divisions.

After he became Rajah, Brooke tried desperately to establish a British colonial presence in the region, but his efforts were continually thwarted as the British were not interested. James Brooke survived in his initial years principally because of British naval support, his co-optation of the local Malay élite, and his use of voluntary Iban warriors to dampen opposition.

The first Rajah died in 1868, leaving an expanded state with new political boundaries to his nephew, Charles. An autocratic ruler, Charles Brooke put the stamp of his strong personality on the map of Sarawak. His son, Charles Vyner, succeeded him in 1917. The final era of the last Rajah came at a time of complex political and economic changes which saw Sarawak being drawn more into the ambit of British colonial control, culminating in the cession of 1946.

1. John Chin, *The Sarawak Chinese* (Kuala Lumpur: Oxford University Press, 1981), p. 107; and Lau Tzy Cheng, *Fung Hsia Miscellaneous Articles* (Singapore: Archipelago Cultural Publisher, 1980), p. 9 (text in Chinese).

2. Initially, I was reluctant to conduct my interviews in the open, due to the lack of privacy, and often requested my informants to adjourn to more conducive surroundings, such as a sitting-room. It gradually dawned upon me that the 'sitting-room' idea is not part of the architectural style of shophouses. A trader usually has a small desk close to the back of his shophouse where he meets his clients and friends. The kitchen is also at the back. The top floor of the shophouse is used as a store and as sleeping quarters by the traders, their families, and their kin. The only free space available is the trader's desk—where I did a number of interviews—and the dining-table in the kitchen, but this was an unsuitable venue, being occupied at different times of the day by housewives preparing their meals, children doing their homework, or as baby-sitting space. People entertain their friends, kin, or clients in the ubiquitous coffee-shop which Fidler has so aptly described as 'everyone's living room'. See Richard Fidler, 'Kanowit: An Overseas Chinese Community in Borneo' (Ph.D. thesis, University of Pennsylvania, 1973), p. 129.

3. James Jackson, *Sarawak, A Geographical Survey of a Developing State* (London: University of London Press, 1968), pp. 15–72.

4. 'North-west Borneo' here includes the present-day state of Sarawak and the nation of Brunei, while 'West Borneo' refers to Kalimantan, taking in Sarawak as well.

5. This tentative dating is based on local evidence collected by Tom Harrisson, a past curator of the Sarawak Museum. Harrisson relied on three sources of information—namely, Idahan texts collected at Madai near Lahad Datu in Sabah, the Brunei royal chronicles, and Iban folklore. See *SG*, 31 August 1968, p. 180.

6. Craig Lockard, 'The Southeast Asian Town in Historical Perspective: A Social History of Kuching, Malaysia, 1820–1970' (Ph.D. thesis, University of Wisconsin, 1973), pp. 6–20.

7. G. William Skinner, *Chinese Society in Thailand, An Analytical History* (Ithaca, New York: Cornell University Press, 1957), p. 37.

8. John Chin, op. cit., p. 17.

9. Diu Mee Kuok, 'The Diffusion of Foochow Settlement in the Sibu–Binatang Area, Central Sarawak, 1901–1970' (MA thesis, University of Hawaii, 1972), p. 43.

10. Skinner, *Chinese Society in Thailand*, p. 40.

11. Lockard, 'The Southeast Asian Town in Historical Perspective', p. 345.

12. John C. Templer (ed.), *The Private Letters of Sir James Brooke, K. C. B., Rajah of Sarawak, Narrating the Events of His Life from 1838 to the Present Time*, Vol. 1 (London: Bentley, 1853), p. 157.

13. Cited by Robert Pringle, *Rajahs and Rebels* (London: Macmillan, 1970), p. 19.

14. Lucas Chin, 'Impact of Trade Ceramic Objects on Some Aspects of Local Culture', *SMJ*, Vol. 25, No. 46 (July–December 1977), pp. 67–9.

15. Erik Jensen, *The Iban and His Religion* (Oxford: Clarendon, 1974), p. 42.

16. See note 12.

17. H. S. Morris, 'The Coastal Melanau', in Victor King (ed.), *Essays on Borneo Societies*, Hull Monographs on South-East Asia 7 (Oxford: Oxford University Press for University of Hull, 1978), p. 39.

18. H. S. Morris, 'How an Old Society was Undermined', *SG*, March 1982, p. 51.

19. Frank LeBar (ed.), *Ethnic Groups of Insular Southeast Asia* (New Haven: Human Relations Area File Press, 1972), p. 170.

20. Sarah Gill, 'Selected Aspects of Sarawak Art' (Ph.D. thesis, Columbia University, 1968), pp. 34–5.

21. Jerome Rousseau, 'The Social Organization of the Baluy Kayan' (Ph.D. thesis, University of Cambridge, 1974), p. 157.

22. Edmund Leach, *Social Science Research in Sarawak* (London: HMSO, 1950), pp. 54–6.

23. The standard conventional texts for these are: S. Baring-Gould and C. A. Bampfylde, *A History of Sarawak under Its Two White Rajahs 1839–1908* (London: Sotheran, 1909; reprinted Singapore: Oxford University Press, 1989), pp. 61–91; and Steven Runciman, *The White Rajahs: A History of Sarawak from 1841 to 1946* (Cambridge: Cambridge University Press, 1960), pp. 45–67.

1
Mining Pioneers in Bau, 1800–1857

THE pioneering efforts of Hakka miners in Kalimantan—a region arbitrarily defined as that area embracing the axis of Sambas, Montrado, Mandor, and Pontianak in Kalimantan, and spilling over into Sarawak—from the mid-eighteenth to the early nineteenth centuries, contributed to the demographic and economic transformation of that part of Borneo (see Map 2 for the distribution of *kongsi* there). The miners' successful adaptation to their new economic and social milieu, and the subsequent transformation of the Bornean landscape that followed, was due primarily to their frontier organization, the *kongsi*.[1] Here, the main features of the *kongsi* will be restated to interpret its significance in a wider regional framework to include Sarawak. The traditions of the mining *kongsi* organization will be examined to show how it became a resilient and autonomous institution. The mining *kongsi* in Bau, like the other *kongsi* in the Kalimantan watershed, struggled to maintain its up-river autonomy. Before reconstructing the story of the Bau mining *kongsi*, it is necessary to explain how *kongsi* were first started in Kalimantan.[2]

Kongsi *Traditions in Kalimantan*

The story of the Hakkas in Kalimantan, who started mining individually and later set up co-operative ventures, merging them into broader brotherhood societies and finally amalgamating with other *kongsi* to form *kongsi* federations, has been recorded by contemporary observers and historians.[3] There are, however, several features of the *kongsi* which warrant attention. The first is the way in which the *kongsi* became an autonomous institution in Kalimantan. The second is the relationship between the *kongsi* and the local people. The third is the changing political balance in the region with the arrival of the Dutch.[4] An examination of the *kongsi* traditions of partnership, brotherhood, and political–economic autonomy has to be understood against this background of political and economic forces, both local and foreign, at work.

There is evidence to suggest that the earliest Chinese miners in Kalimantan were invited there around the mid-eighteenth century

Map 2 *Kongsi* in West Borneo and Sarawak

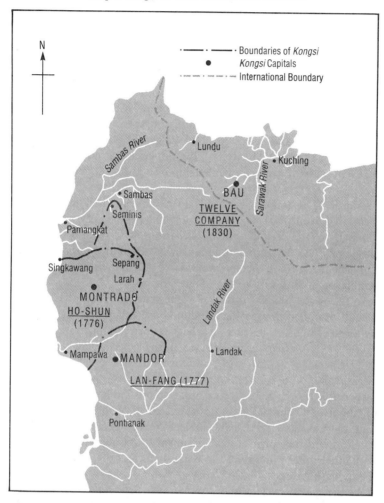

by the Malay sultans to mine gold ore. Tombstones of Chinese miners dating back to 1745 have been found in Mampawa and Mandor.[5] Around 1760, the Panembahan of Mampawa, perhaps influenced by the success of the Sultan of Bangka in using Chinese tin-mining labour, asked twenty Chinese from Brunei to dig for gold ore in the Doeri valley.[6]

The miners, who were predominantly Hakka, formed the first of their mining partnerships, called *shan-sha*[7] (mountain of sand) or *parit* (mine) in this pioneering period between 1743 and 1745. These mines, run by between ten and twenty-five men, were subject to the control of the Malay sultans on the coast. The sultans looked upon the Chinese as a source of income, and monopolized and regulated their flow of supplies, tools, provisions, and opium. The Hakka miners, on their own initiative, raised their own capital, and protected their own interests through mutual support for each other. By 1764, the Hakkas had formed twelve mining unions in Larah and twenty-four in Montrado.[8]

As more mines were established, and as more newcomers arrived, the rudimentary form of partnership among the Hakkas was enlarged and modified to take on a political role, a form of brotherhood government called the *hui*.[9] This critical development started around 1763. The *hui*, with an average membership of between 50 and 150 men, were powerful enough to challenge the sovereignty of their Malay overlords. The miners were able to circumvent the unfavourable terms of forced trade which the sultans had previously imposed on them. By 1770, irregular and partial tribute was paid to the Malay suzerains.[10] As the mining unions grew in strength, Malay overlordship declined, until the intervention of the Dutch restored some semblance of authority to the Malay sultans.

In the 1770s, the word '*kongsi*' came into use. It emphasized the idea of extended mining partnership and brotherhood, symbolized by a *kongsi*-house.[11] Each *kongsi* had its own administration, again called '*kongsi*'. *Kongsi* office-bearers were elected by members, and daily financial matters were administered by several appointed clerks, called '*ts'ai-ku*'. Out on the mine, overseers—*huo-chang*—were responsible for the task of procuring gold ore and for the supervision of labour. The general length of service of these office-holders was four months, and they were subject to dismissal during general meetings. In the spirit of egalitarianism, the leaders lived together with other mine workers in the *kongsi*-hall, part of which contained the living quarters.

As the *kongsi* became more numerous, they sought to combine together to form federations. In 1776, fourteen *kongsi* in Montrado agreed to unite under a federation, the Ho-shun *kongsi* (meaning the Federation of Great Peace and Harmony), thus broadening the concept of extended brotherhood and partnership.[12] Still more *kongsi* in neighbouring Larah were to seek membership in the Ho-shun *kongsi*, with its central headquarters or *tang* (hill) in Montrado.

To the south, on the Mandor field, an enterprising individual, Lo Fang Po, and his Chia-ying Hakka followers gained control of the Tapu Hakka *kongsi* at Shan-hsin in the mid-1770s. Under Lo's leadership, this mixed group of Chia-ying and Tapu Hakkas went on to take up mining around Mandor, creating a federation known as the Lan-fang (Orchid Fragrance) *kongsi*. It was also known as *ta-tsung-chieh* (presidential system or 'republic'), and Lo Fang Po, its leader, was acclaimed *ta-tsung-chang* (great president). The Lan-fang *kongsi* had an elaborate system of government. The headman or *ta-ko* (eldest brother) presided over the whole *kongsi* like a head of state, a position reserved for the headman of Mandor. Assistant headmen, *fu-t'ou-jen wei-ko* or *lo-tai* (elder brother) were elected to district-level leadership.[13] Lo Fang Po divided the areas under his jurisdiction into administrative districts called provinces, which were then subdivided into prefectures. The prefectures were further subdivided into counties, under the charge of captains. Magistrates and captains served in various administrative and judicial offices, responsible for matters such as toll collection, tax levying, and immigration.

At the beginning of the nineteenth century, the two *kongsi* federations, the Ho-shun in Montrado and the Lan-fang in Mandor, were engaged in both internecine warfare and fighting against external enemies, the Malay sultans and their Dayak supporters, and the Dutch. Internal *kongsi* fighting reduced the original Ho-shun federation to three members only, led by the Ta-kang (Great Creek) *kongsi*; this faction was opposed by the San-tiao-kou (Three Watercourse) *kongsi*, which had left the grouping to operate at Sepang. The Lan-fang *kongsi* federation under Lo Fang Po, which had abandoned plans to annex Montrado, strengthened itself in the south-east, at the expense of the Sultan of Landak. However, by the 1830s, the initiative in warfare had returned to the Sultan and his Dayak followers. Intermittent warfare between the Lan-fang *kongsi* and the Sultan of Landak and his Dayak followers continued into the following decade, with the Sultan of Pontianak acting as arbitrator in the dispute in 1842 and again in 1846. The Lan-fang *kongsi* was weakened by the fighting and went into decline. It was not strong enough to resist the signing of a non-aggression pact with the Dutch in 1848. In return for an undertaking of non-interference in the internal affairs of the *kongsi* by the Dutch, the *kongsi* agreed to appoint as headman one Liu Ah Shin, who was known to be sympathetic to the Dutch.[14]

The crumbling Ho-shun *kongsi* federation in Montrado faced Dutch pressure as early as 1822. The Dutch attempted to circumscribe the autonomy of the *kongsi* by imposing trade restrictions, limiting the flow of overseas trade through the Dutch-controlled ports of Pamangkat and Sambas, and imposing capitation and departure taxes. These conditions were not strictly enforced because Dutch attention was continually distracted by what was occurring in other parts of the archipelago. Nevertheless, the Ta-kang *kongsi*, opposed to these conditions in principle, waged war against the Malay suzerain, the Sultan of Sambas, in 1848.[15] In 1850, the Dutch returned to the area with the intention of subduing the Ta-kang *kongsi* by helping the Sultan of Sambas. In response, the Ta-kang leaders strove to drive the Dutch out of the Sambas area. To do this, they had to cross rival San-tiao-kou territory, which they did in October 1850, causing an overspill of San-tiao-kou refugees into Bau, Sarawak. Dutch reinforcements soon arrived, blockaded the coast and, by the end of the year, the Ta-kang *kongsi* had capitulated.

The preceding discussion on the *kongsi* in Kalimantan illustrates the growth and the political and economic independence of the *kongsi*, and places in a regional context what was to occur in Bau from the beginning of the nineteenth century onwards. The *kongsi* federations which emerged in Kalimantan in the mid-eighteenth century were locked in rivalry with one another, while also being engaged in constant warfare against the Malay sultans and their Dayak allies, and

Map 3 Bau Mining District

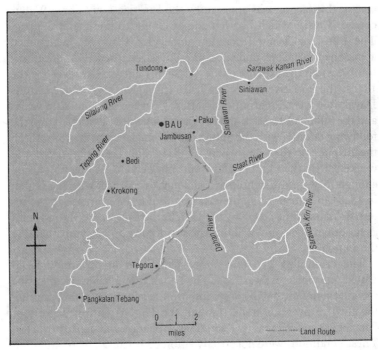

then the Dutch, in an effort to preserve their independence. This state of affairs continued until the eventual collapse of the *kongsi* in the mid-nineteenth century. When the miners across the watershed in Bau, Sarawak, established their *kongsi* around the turn of the century, the geopolitical adversities they faced in attempting to preserve their autonomy were not unlike those experienced previously by the Kalimantan *kongsi* miners. This time, the Bau *kongsi* miners had to contend with the indigenous people and with an external power in the person of James Brooke.

The Origins and Growth of the 'Twelve Company' in the Bau Mining District

It is not known when the Hakkas first began to mine gold in Bau.[16] Like the miners in Kalimantan, the Bau Hakkas started off individually and in small groups. When antimony ore became a valuable commodity in the Singapore market in the 1820s, Hakka miners were known to be mining the mineral ore in Bau.[17] They continued to do so until the civil war between the local Malays and their Dayak allies, on the one hand, and Pengiran Muda Hashim, the Brunei Sultan's representative in the Sarawak River basin, on the other, put a temporary halt to their activities. Local oral tradition dates the founding of a permanent settlement to around 1830.[18] Oral recollections claim that the particular *kongsi* established was the San-tiao-kou *kongsi*, a claim which corroborates written European sources.

After antimony ore, the pioneers then started mining for gold at Lumabau, not far from present-day Bau. The miners were under the leadership of Liew Shan Pang, who headed a *kongsi* with eleven other leaders—hence the *kongsi*'s popular name, 'Twelve Company'. The mining operations of the *kongsi* extended from Lumabau to Bedi, Paku, and Tundong. (Map 3 shows the distribution of settlements.) The *kongsi* had a small beginning. The original mining enclave could have numbered no more than 200 persons, but population growth augmented by watershed migrations could have increased the size of the community to approximately 600 by 1848.[19] A significant increase in the size of the mining community occurred in 1850 when an estimated 3,000 members of the San-tiao-kou *kongsi*, from Sepang in Kalimantan, crossed over into the Bau district to escape from the ravages of the Ta-kang *kongsi*–Dutch war.[20] The arrival of the newcomers caught James Brooke, Rajah of down-river Kuching since 1841, by surprise. The Rajah's immediate concern was whether he could persuade the new arrivals to reside near Kuching, or allow them to swell the ranks of the Bau *kongsi*. In a journal entry, James Brooke wrote confidently:

Sarawak flourishes, and an influx of a large body of Chinese promises well for the future, and the fact of these immigrants being agriculturists and having wives ensures the quiet of the country.... As the Santiqu [San-tiao-kou] are now in distress and sufficiently humble, it will not be a difficult task to reduce it to obedience, and to establish a fixed system of government.[21]

Any notions the Rajah had of bringing the new arrivals under his direct aegis failed. He attempted to resettle the newcomers down-river at Muara Tebas, a tributary of the Sarawak River, and at Santubong, the entrance of the main river, where they would be easily controlled from Kuching. Evidently, the majority of the San-tiao-kou members preferred to move up-river to Bau, coming under the jurisdiction of the *kongsi* there.[22]

The Bau mining settlements were distributed in an area between the confluences of the Silalang, Tepong, and Siniawan Rivers, tributaries of the Sarawak River. Tundong, situated along the Silalang River, was a mixed mining–agricultural station. Some residents were engaged in antimony mining at Buso, and in digging for gold ore at Bau, while others were market gardeners, planting vegetables and supplying food to the miners. Between Siniawan and Tundong, there were fifteen settlements.[23] Siniawan itself was a mixed trading–agricultural centre, with a population of some 300 Hakkas serving the surrounding clientele of Land Dayaks and Hakka and Malay gold-workers.[24] The 'gold district' was located between the Siniawan and Lower Tepong Rivers. The miners built a network of ten roads running from Siniawan to Bau, wide and hard enough to be ridden over by horses.[25] Bau was not accessible by river from Siniawan, but could be reached by road from Tundong and Siniawan. Bau, as the headquarters of the mining district, had 100 shops[26] and as many thatched huts.[27] The mining district took on the character of a Chinese pioneering frontier, with mines and farms and well-ordered settlements served by shops and roads. The physical and political landscape of the Bau district was thus transformed.

In 1856, Spenser St. John, private secretary to the first Rajah, made a tour of the extra-Kuching settlements. He started off with the rice-fields and vegetable gardens at Sungei Tengah, a tributary of the Sarawak River, 6 miles south-west of Kuching, and proceeded up-river to the small township of Siniawan. From Siniawan to Bau, there was a constant succession of reservoirs and gold mines.[28] There was a road built by the pioneers from Bau along the boundary of the watershed with Kalimantan, which St. John described as an 'admirable path' for the purposes of 'facilitating intercourse with Sambas'.²⁹

The mixed mining–agricultural settlements were self-sufficient in their needs, particularly in trade and food supplies, and they were oriented towards *kongsi* on the other side of the watershed, most possibly towards the San-tiao-kou *kongsi*, especially since approximately 3,000 of its members had arrived in Bau in 1850. The Bau district, as an autonomous zone, had little to do with down-river Kuching, only 16 miles away. James Brooke regarded the Bau Hakkas as hostile pioneers, while the Kuching-based Chinese, mostly Hokkien and Teochiu traders, were a welcomed mercantile community.³⁰ As the Hakka miners had, without any outside help, pioneered and opened up the Bau frontier, they naturally laid strong claim to control of the district. Just as the *kongsi* across the watershed in Kalimantan had enjoyed political autonomy, so did the Bau *kongsi* strive to achieve a similar objective.

Early Political, Economic, and Social Conditions of the 'Twelve Company' in Bau

The *kongsi* formed the basis of the miners' political and social organization. The *kongsi* was an instrument of self-government and looked after its members' economic and social needs.

Contemporary observations and oral accounts confirmed the autonomous status of the *kongsi* (more commonly known as the 'Twelve Company' or San-tiao-kou *kongsi*). Ludvig Helms, a Danish merchant in Kuching, noted that the *kongsi* 'was governing [the miners], electing its own magistrates, inflicting the punishment of death, and in a word, was independent of the Rajah's government'.³¹ The *kongsi* made its own guns and weapons,³² and minted its own copper coins.³³ It administered the scattered Chinese communities near by.³⁴ It flew its own flags, the flag-poles of which remain to this day. According to local tradition, the *kongsi* flags were raised every morning to symbolize the political supremacy of the *kongsi*, and offenders convicted of *kongsi* crimes which carried the death penalty—crimes such as adultery—were executed in front of the flags.³⁵

The self-government of the *kongsi* was viewed unfavourably by observers who were inclined to believe that only James Brooke's government in Kuching was strong and proper. In retrospect, Helms was of this view:

... the Chinese had reason to be happy with their lot; they lived under a government which, as long as they consumed and paid taxes upon their consumption, left them very much to themselves; indeed they were not sufficiently governed and looked after.... Had the Chinese in Sarawak (including Bau), which by this time [1856] numbered over 4,000, felt the hand of a strong government (James Brooke) upon them they would not have risen in rebellion.[36]

Another writer, Rodney Mundy, while praising the Chinese for their industry, said:

The Chinese have many good points; they are active, industrious and commercial; and when we consider their ignorance, and the badness of the government under which they have lived, deprived of trade, and subject to all the evils of extortion and monopoly, we are apt to give them such credit for the good qualities they display.[37]

From these two reports, it can be seen that the *kongsi* was quite independent of James Brooke. The democratic features of the *kongsi* were briefly noted, and seemed to be similar to what Wang Tai Peng has said about the West Bornean *kongsi* in Kalimantan, in the openness of its government, and in the notions of extended partnership and brotherhood among the miners. The Bau *kongsi* was 'governed by officers selected by the multitude, a common workman may rise suddenly to be their leader; they generally look at the business character of the men put up for selection'.[38]

The symbol of government, as well as the focus of community life, like elsewhere in Kalimantan, was the *kongsi*-house, situated in the valley of Bau, flanked on two sides by black-looking steep hills. It was a substantial structure, built of *belian* (ironwood) posts and sturdy planks, and roofed with *belian* shingles.[39] James Brooke, on a visit to Bau, gave a description of the Bau *kongsi*-hall: 'The temple is in the centre; on each side is a room where the workmen sleep, and above these again the stores are quickly stowed away, on floors beneath the roof. Without is an open shed, about the ground which the kongsi-folk and labourers sit and eat, and drink tea.'[40] As an institution, the *kongsi* served to fulfil the spiritual needs of the pioneers. Facing the entrance was the altar in honour of the *Tai Pek Kong*, the deity for pioneering Chinese. The deity was probably derived from Tu Ti (the Earth God) but, overseas, it symbolized the spirit of the pioneers.[41] Tablets honouring those who had given worthy service to the *kongsi* were hung on the sides of the altar. This was where new members were initiated. The memory of pioneers was honoured by those who came after them. There is a small temple in Bau which honours

Liew Shan Pang, the leader of the *kongsi*. Workmen and their elected leaders lived together in the *kongsi*-house to emphasize unity and egalitarianism.

The Hakkas employed a mining technology which was more advanced than the panning techniques used by the local Malays and Land Dayaks. They constructed sluices, reservoirs, and canals, giving the landscape a distinctly Chinese appearance. The following passage describes a typical Hakka mining operation:

They dammed up the end of the valley at the back of the *kunsi*'s house, thus forming a large reservoir of water, a quarter of a mile in length. The dam was neatly constructed, completely faced with wood towards the water, and partially on the outside to resist the heavy rains. A ditch, about four feet broad, was cut from the reservoir towards the ground which the overlooker of the company had selected as the spot likely to produce a good yield of gold, and a well-made sluice-gate was constructed in the dam to supply the ditch with as much water as might be required; minor sluice-gates to the main ditch enabled the smaller ones also to reduce supplies of water. After this was ready, the sluice-gates were opened and the earth in its neighbourhood thrown into the ditch and the rushing water carried off the mud and sand, and allowed the particles of gold to sink to the bottom. After 3 or 4 months, they cleaned out the ditch, and carefully washed the residue which generally yielded them sufficient to make a tolerable division among the workmen after all the expenses had been paid.[42]

The economic organization of the *kongsi*, in sharing work and profits, was well suited for mining activity of this scale which involved the construction of many dams and water-gates, and a long wait of three to four months before the residual gold ore could be cleaned out. Where the collective efforts of the *kongsi* were not required, a less sophisticated way of obtaining gold by panning was employed. At Jaong, behind Sarambo Mountain, situated between Bau and Buso, individual Chinese miners formed pits of about 10 yards square, in which the soil was first liquefied and then carried by the flow of the liquid into a pond where the residual metal ore collected. The upper layer of the liquefied matter was drained off and the residue was panned by hand.[43] This work was undertaken by women. An account written in 1865 described how women performed this task: 'They had crude cradles for gold washing; when the jars were filled, they scraped up a mass of mud from the bottom of the river, and commenced twirling their cradles in the usual manner.'[44]

That women were involved in the simpler process of gold-panning was significant. It emphasized the work ethic of the mining community, but with women performing the less physically demanding tasks. Many of the Hakka men had Land Dayak spouses. Over an extended period of time, a mixed Chinese–Land Dayak community had emerged in Bau, with kinship links traced back to Sambas.[45] There were few obstacles to Chinese–Land Dayak marital unions. There was no religious barrier to mixed marriages. The eating of pork, drinking of alcohol, and a propensity for gambling were social characteristics shared alike by the Chinese and the Land Dayaks. The fact that the Hakka miners treated their native wives well could have made these cross-cultural marriages a stable feature of *kongsi* life.[46] For the women, marriage to Chinese men was believed to lead to a better standard of living materially, given the reputed industry of the miners. The children of these marriages were invariably brought up as Chinese. This could have been due to the influence of the *kongsi* which, as the focus of political power, helped to preserve Chinese cultural and social values.

Pioneering conditions were harsh, and the men and their families lived simply and had few pleasures in life to compensate for their hard work. It was graphically said: 'A Chinaman is willing to work on condition that he is well fed; he wants his four or five meals a day, consisting of rice, vegetables, and pork. He must have his tea, tobacco, opium and *samshu*, a spirit distilled from rice, and when he has money, he must gamble.'[47]

Opium, in particular, had therapeutic value, especially for those who needed to rest after a hard day's work, and it served as a pastime as well. Charles Brooke, who visited the Marup gold mines in the Batang Lupar in 1863, recorded this observation of an old Chinese man taking opium. What he saw could easily have taken place in the Bau mines:

> We cooked and dined in a tumble-down place, but it was better than a confined boat. Our dinner consisted of stew and vegetables, and when it was over, while sipping alcohol and smoking, I watched an old Chinaman after his day's labour, crawl deliberately to his sleeping bench; then kicking off his shoes, he entered his curtains, pulled a small saucer lamp from one corner, lighted it, then opened an account book, in which he wrote and made some calculations very silently. Then, he reduced his lamp flame, heaved a heavy sigh, as much to say, 'another day's work over', and now for that short period of Elysium before the body sinks into forgetfulness. He loaded his pipe with opium, sank his head on the pillow, and lying on one side, lighted it, drew for about three minutes that soothing and delicious narcotic. His pipe then fell, a breath extinguished the light, and he had already gone into some paradise of a Chinese creation.[48]

Generally, the Hakka miners appeared 'comfortable and peaceful',[49] and well accommodated to their Land Dayak neighbours with whom they intermarried. It was not uncommon for the miners to take the families and kin of their Land Dayak wives under their care.[50] The mixed offspring of the Chinese and the Land Dayaks were described as a 'sturdy race';[51] and the women were praised for their good looks.[52]

A strong feature of the *kongsi* was its independence. As a frontier organization, the *kongsi* served the needs of the miners in their new milieu. It enjoyed political autonomy, and the socio-economic organization of the *kongsi* enabled the members to carry out their mining activities efficiently. The *kongsi* gradually became a self-sustained community, well integrated with its environment. Like the *kongsi* federations in the rest of West Borneo struggling to keep their independence, the Bau *kongsi* had to contend with James Brooke and his native allies.

Kongsi–Brooke Rivalry

The autonomous existence of the *kongsi* was ultimately challenged by James Brooke. It was by no means certain that James Brooke could claim uncontested sovereignty over Bau. Conflicting claims to dominion were to arise. The Kuching-based Rajah was to try to impose his will on Bau, extend his ideas on criminal jurisdiction, and levy 'taxes' on opium consumed by the Bau residents. James Brooke's notions of sovereignty in Borneo developed soon after he had acquired the Raj from Pengiran Muda Hashim, the Brunei Sultan's representative in Kuching, through a classic exhibition of gunboat diplomacy.[53] Brooke's idea of sovereignty rested on control over territory and free trade.[54] In the local Bornean context, where Malays assumed politico-economic power at strategic points along rivers and coasts, sovereignty was exercised by control over people, through forced trade and the exaction of tributes, not by control over territory.[55] In either concept of sovereignty, Kuching had no direct control over Bau.

The local Bau perspective claims that the *kongsi* first negotiated its agreements with Pengiran Muda Hashim[56] and that these rights were not transferred to James Brooke. Oral tradition also claims that the boundaries between the respective domains of James Brooke and the *kongsi* were clearly delimited.[57] Though James Brooke had established himself in Kuching in 1841, he was not content for long to remain merely within those boundaries, and was to enforce and spread his overlordship over other river basins through a series of bloody battles against Iban 'pirates' between 1841 and 1849. The autonomous Bau mining district was not left out of Brooke's plans for expansion: for a start, he would have found the idea of the *kongsi* running its own affairs anathema, especially when it was in control of a mineral-rich region. In any case, the *kongsi*, which had established itself even before James Brooke first set foot in Borneo, in all likelihood deemed him an intruder, and considered his claim to be a down-river authority with control over an up-river settlement as not unchallengeable. Conflict was therefore inevitable. A recurrent pattern in other parts of West Borneo in the latter part of the eighteenth century was that when *kongsi*, usually situated up-river, became strong enough, they challenged the authority of down-river centres. A similar geopolitical conflict was to take place in Sarawak between the up-river Bau *kongsi* and the downriver-based White Rajah.

In the initial years of the Raj, the equation of power between Kuching and Bau was roughly equal. James Brooke was of course anxious to use force to make the *kongsi* capitulate to his demands. The Rajah tried to force his writ upon the *kongsi* as early as 1842. The opportunity to do so presented itself when a legal question arose over an application by a rival *kongsi* to James Brooke for the right to mine gold in Bau. This application was opposed by the 'Twelve Company'. James Brooke successfully used a display of armed force to persuade the 'Twelve Company' to change its mind. The Rajah himself admitted that in Bau, 'here was evidently an *imperium in imperio*, which [will foresee] future trouble'.[58] Several more occasions provided James Brooke with opportunities for attempting to circumscribe the autonomy of the *kongsi*.

In 1852, Rajah Brooke suggested a sharing of political power and joint Bau–Kuching administration of the Bau district. The Rajah first called for a conference with three *kongsi* leaders and thrust these demands upon them:

> That the kongsi was not to aim at governing, nor to receive revenue, but was frankly, freely and unreservedly, to acknowledge the authority of the Government of Sarawak. [The miners] were to appoint a Captain China whilst I [James Brooke] was to appoint a Captain Inglis, who were jointly to superintend the affairs and proceedings of the company in the interior. . . . The kongsi was not to decide any cases of dispute or crime, excepting among their own people, and then only for misdemeanours of a light description. . . . The kongsi was not to take Dayak lands, nor to open up new *parits* without permission.[59]

In saying this, the Rajah himself had implicitly acknowledged the autonomous existence of the Bau *kongsi*, a state of affairs he hoped to remedy in his favour. It is not known how long the joint 'captain' arrangement lasted, whether it worked satisfactorily from the standpoints of both parties, or whether the *kongsi* ever accepted such an arrangement in the first place. The author would venture to conjecture that such a proposed arrangement did not work, if indeed it was ever put into practice, for a year later, the Rajah had to send an armed force up-river to apprehend a wanted man. This meant that James Brooke did not treat the matter lightly when the *kongsi* at first refused to hand over the wanted person. Anticipating resistance from the *kongsi*, extra precautions were taken in Kuching, such as keeping watch over public buildings, and an armed armada of small canoes, commanded by Charles Johnson (Brooke), went up-river.[60] According to the future second Rajah's account, the *kongsi* acquiesced to Kuching's demands in giving up the fugitive.[61] As 'compensation', the *kongsi* leaders were required to build a fort at Belidah, bearing the cost of arms, ammunition, and the wages of the men posted at the fort.

The use of force, a proposal to share political power, and the claim to criminal jurisdiction were examples of Rajah Brooke putting pressure on the 'Twelve Company'. But James Brooke did appear to have been able to flex his political muscles over Bau if the occasion so demanded. Though the *kongsi* occasionally yielded to pressure exerted on it by James Brooke, it was nevertheless able to continue to enjoy economic control over its resources and trade.

Bau did not export its gold through Kuching. The gold ore was sent to Sambas four times a year, and the 'Twelve Company' was reported to be 'in constant communication with Montrado and Sambas'.[62] The expansive economic network of the Bau *kongsi* was linked to the *kongsi* federations in Kalimantan. The 'Twelve Company' was able to avoid using Kuching as an entrepôt even though the Sarawak River was the only waterway connecting Bau to the coast, because of land routes across the watershed which linked the various *kongsi* together, and because it preferred to deal with neighbouring *kongsi*. The trade of Kuching was in the hands of Hokkien and Teochiu merchants. There was mutual mistrust, antagonism, and lack of contact between the urban Kuching Hokkiens and Teochius, and the rural Bau Hakkas. The Hokkien and Teochiu traders in Kuching would have seen the Hakkas in Bau as bitter rivals, especially when they had no share of the lucrative trade in gold-ore exports. At a time when the Hakkas themselves were in control of the mining and trade in gold, it was unlikely that they would countenance the idea of sending the mineral ore through the network of Kuching Hokkien and Teochiu traders.

The Bau *kongsi* paid an opium 'tax'. It has been commonly thought that payments made by the *kongsi* to Rajah Brooke constituted a form of tributary payment in tacit recognition of Brooke's sovereignty. However, there is evidence to suggest that the payment was a form of 'tax' on the opium bought for consumption by the *kongsi*.[63] James Brooke arbitrarily levied an annual payment of 60 *tahil* of gold on the miners based on a given population size. When the quantity of opium recorded as consumed failed to keep up with the expanding population of the *kongsi*, the Rajah came to the conclusion that much opium was being smuggled in from Sambas. He came to regard this as an infringement of his 'tax' right, and unilaterally raised the 'tax'— a move opposed by the *kongsi*. This raising of the opium 'tax' has been commonly cited by Brooke officials, observers, and historians as being the cause of the *kongsi*–Brooke war. However, this is a rather weak explanation for it ignores the long-term differences between the *kongsi* and James Brooke.

Other reasons have been given for the outbreak of war. The proceedings of the Commission of Inquiry held in Singapore in 1854, which investigated James Brooke's conduct against Iban 'pirates', were said to have given encouragement to the miners in Bau, in view of the unfavourable publicity for Brooke.[64] 'Secret society' (triad society or *Tien Ti Hueh*) connections, aimed at overthrowing the Ching Dynasty in China, have also been suggested as being responsible for encouraging the Bau *kongsi*.[65] However, that *kongsi* had little to do with events overseas, in China or elsewhere, and most of its dealings were with neighbouring *kongsi* across the watershed. If there was any external influence on Bau, it would have come from Kalimantan. The anti-British feelings after the Canton incidents of October 1856 have been cited as a factor to explain the outbreak of war between James Brooke and the *kongsi*.[66] If that was a plausible reason, then it would be expected that anti-Brooke sentiments would have been more widespread, rather than being confined to Bau alone. The Kuching Chinese were not sympathetic towards the 'Twelve Company', and the Hakka miners in Marup, Batang Lupar, whose 'Fifteen Company' organization was as formidable as Bau's, did not take part in the conflict. This would tend to suggest that the differences between Rajah Brooke and the *kongsi* were of a localized nature, a contest for ascendancy between two sections of a river. Such a pattern of conflict was to be a familiar theme during the reign of the

Brookes in Sarawak. Oral tradition also claims that economic rivalry between the *kongsi* and the newly formed European concessionaire, the Borneo Company, established in 1856, was a significant factor in the *kongsi*'s conflict with the Rajah.[67] There are indeed strong reasons in support of this claim, as events after 1857 proved that the miners' fears of their *kongsi* being undermined by the Borneo Company were well justified.

The pioneering Hakkas had established a unique frontier organization, the *kongsi*, which enabled them to adjust to their new environment. The *kongsi*, with its emphasis on democratic principles and self-government, successfully transformed the demographic landscape of Bau into a Hakka mining district. The principles espoused by the *kongsi*—those of sharing work, resources, and profits—were suited to the collaborative efforts of gold mining. Bau, as the headquarters of the *kongsi*, became a regional centre which rivalled downriver Kuching, ruled by James Brooke.

---

1. See John Chin, *The Sarawak Chinese* (Kuala Lumpur: Oxford University Press, 1981), Chapter 2; James Jackson, *Chinese in West Borneo Goldfields: A Study in Cultural Geography*, Occasional Paper in Geography, No. 15 (Hull: University of Hull, 1970); Chang Tsuen-Kung, 'Historical Geography of Chinese Settlement in the Malay Archipelago' (Ph.D. thesis, University of Nebraska, 1954), Chapters 4 and 5; Barbara Ward, 'A Hakka *Kongsi* in Borneo', *Journal of Oriental Studies*, Vol. 1, No. 2 (July 1954), pp. 358–70.

2. This section relies heavily upon ideas developed by Wang Tai Peng, 'The Chinese Republic in West Borneo from the Latter Part of the 18th Century to the Middle of the 19th Century' (MA thesis, Australian National University, 1977); Craig Lockard, 'The Southeast Asian Town in Historical Perspective: A Social History of Kuching, Malaysia, 1820–1970' (Ph.D. thesis, University of Wisconsin, 1973), Chapters 1 and 2; and Jackson, *Chinese in West Borneo Goldfields*, Chapters 2 and 3.

3. Barbara Ward, op. cit., p. 351; Spenser St. John, *Life in the Forests of the Far East*, Vol. 2 (London: Smith Elder and Co., 1862; reprinted Singapore: Oxford University Press, 1986), p. 334; Jackson, *Chinese in West Borneo Goldfields*, p. 61; Wang Tai Peng, 'The Chinese Republic in West Borneo', pp. 55–6; and Wang Tai Peng, 'The Word Kongsi: A Note', *JMBRAS*, Vol. 3, No. 1 (1979), p. 102.

4. Lockard, 'The Southeast Asian Town in Historical Perspective', pp. 21–32.

5. Jackson, *Chinese in West Borneo Goldfields*, p. 20.

6. Lo Hsiang Lin, *A Historical Survey of the Lan Fang Presidential System in Western Borneo by Lo Fang-Pai and Other Overseas Chinese* (Hong Kong: Chinese Cultural Institute, 1960), cited by John Chin, op. cit., p. 13.

7. Wang Tai Peng, 'The Chinese Republic in West Borneo', p. 71.

8. Ibid., p. 55.

9. Ibid., p. 71.

10. Jackson, *Chinese in West Borneo Goldfields*, p. 53.

11. The *kongsi*-house or *tang* was a distinctive feature of the mining landscape. It was often a substantial structure built of *belian* surrounded by a rectangular earthen wall and heavily fortified. The *kongsi*-house was often centrally located, and its presence symbolized the institution of the *kongsi*. In the *kongsi*-house were the quarters of its members and rooms for storing common property, tools, weapons, provisions, funds, and gold.

12. Wang Tai Peng, 'The Chinese Republic in West Borneo', p. 77.

13. Ibid., pp. 9–10; and Barbara Ward, op. cit., p. 14.

14. Barbara Ward, op. cit., p. 14.

15. Henry Keppel, *A Visit to the Indian Archipelago in H.M. Ship Maeander with Portions of the Private Journal of Sir James Brooke, K.C.B.*, Vol. 2 (London: Bentley, 1853), pp. 51–2.

16. A geologist, W. E. Wilford, conjectured that, early in the nineteenth century, Chinese miners were obtaining gold by panning and sluicing methods near Pangkalan Tebang, just immediately north of the Indonesian border. See W. E. Wilford, *The Geology and Mineral Resources of the Kuching–Lundu Area, West Sarawak, including the Bau Mining District* (Kuching: Government Printing Office, 1955), p. 155.

17. Spenser St. John, *Life in the Forests of the Far East*, Vol. 2, p. 321.

18. Liew Peck Kwee, *A History of the Hoppo Chinese with Special Reference to Sarawak* (Singapore: Tung Yet Publishing House, 1978), p. 41 (text in Chinese).

19. Spenser St. John, *Life in the Forests of the Far East*, Vol. 2, p. 323. There was another Hakka mining enclave at Marup, in the upper reaches of the Batang Lupar in the Second Division. During the Ta-kang *kongsi*–Dutch war of 1850, when San-tiao-kou refugees fled into Bau, a group entered the Simanggang area. The *kongsi* that was formed in Marup called itself the 'Fifteen Company' (after its fifteen leaders), and it had 300 workers. See Liew, *A History of the Hoppo Chinese*, pp. 46–7.

20. Spenser St. John, *The Life of Sir James Brooke, Rajah of Sarawak* (London: William Blackwood & Sons, 1879), p. 225.

21. Keppel, *A Visit to the Indian Archipelago in H.M. Ship Maeander*, Vol. 2, pp. 51–2.

22. However, by the beginning of 1856, nearly 500 of the new settlers were established at Sungei Tengah, a tributary of the Sarawak River, 6 miles south-west of Kuching. See Harriette McDougall, *Letters from Sarawak*, reprint edition (London: Wheldon and Wesley, 1924), pp. 117–18.

23. Keppel, *A Visit to the Indian Archipelago in H.M. Ship Maeander*, Vol. 2 , p. 67.

24. Ibid., p. 152.

25. Spenser St. John, *Life in the Forests of the Far East*, Vol. 2, p. 332.

26. Ibid.

27. Keppel, *A Visit to the Indian Archipelago in H.M. Ship Maeander*, Vol. 2, p. 27.

28. Spenser St. John, *Life in the Forests of the Far East*, Vol. 2, p. 335.

29. Ibid.

30. Gertrude L. Jacob, *The Raja of Sarawak: An Account of Sir James Brooke, K.C.B., LL.D., Given Chiefly through Letters and Journals*, Vol. 1 (London: Macmillan, 1876), p. 236.

31. Ludvig Verner Helms, *Pioneering in the Far East, and Journeys to California in 1849, and to the White Sea in 1878* (London: W. H. Allen, 1882), pp. 164–5.

32. Charles Brooke, *Ten Years in Sarawak*, Vol. 1 (London: Tinsley, 1866), p. 28. A Malay miner, in 1907, discovered a gun with a faint inscription, 'General of the Sam Tiow Kew (San-tiao-kou) kongsi'. See *SG*, 4 January 1907, p. 10.

33. Liew Peck Kwee, 'The Relationship between the Bau Kongsi and the Brooke

MINING PIONEERS IN BAU   35

Government', in Teochew Association, Kuching, *Centenary Volume* (Kuching: Teochew Association, 1965), p. 171 (text in Chinese).

34. There is evidence to suggest that the influence of the *kongsi* spread as far as Lundu, which caused James Brooke to feel resentful when he learned of it. In a letter written by the Rajah to an officer, Charles Grant, he said: 'You speak of two kongsi— and the Bow [Bau] kongsi are building at Lundu—this matter needs attention before any others and if it be true, immediately stop it as it is done without leave.' (James Brooke to Charles Grant, 18 May 1854, MSS Pac. 590, Vol. 4, Rhodes House, Oxford.)

35. *BB*, 19 June 1982.
36. Helms, op. cit., p. 159.
37. Rodney Mundy, *Narrative of Events in Borneo and Celebes, down to the Occupation of Labuan: From the Journals of James Brooke Esq.*, Vol. 1 (London: John Murray, 1848), p. 286.
38. Spenser St. John, *Life in the Forests of the Far East*, Vol. 2, p. 334.
39. Ibid., p. 326.
40. Keppel, *A Visit to the Indian Archipelago in H.M. Ship Maeander*, Vol. 2, p. 67.
41. Victor Purcell, *The Chinese in Malaya* (Kuala Lumpur: Oxford University Press, 1967), p. 125.
42. Spenser St. John, *Life in the Forests of the Far East*, Vol. 2, p. 326.
43. Mundy, op. cit., pp. 385–6.
44. Frederick Boyle, *Adventures among the Dayaks of Borneo* (London: Hurst and Blackett, 1865), pp. 69–70.
45. Spenser St. John, *Life in the Forests of the Far East*, Vol. 2, p. 334.
46. Boyle, op. cit., p. 71
47. Helms, op. cit., p. 157.
48. Charles Brooke, *Ten Years in Sarawak*, Vol. 1, pp. 160–1.
49. Mundy, op. cit., Vol. 1, p. 372.
50. Henry Ling Roth, *The Natives of Sarawak and British North Borneo*, Vol. 1 (London: Truslove and Hanson, 1896), p. 124.
51. McDougall, *Letters from Sarawak*, p. 133.
52. Spenser St. John, *Life in the Forests of the Far East*, Vol. 1, p. 153.
53. Graham Saunders, 'James Brooke and Asian Government', *BMJ*, Vol. 3, No. 1 (1973), p. 114.
54. Robert Pringle, *Rajahs and Rebels* (London: Macmillan, 1970), p. 71.
55. The situation in the Malay peninsula was similar. See J. M. Gullick, *Indigenous Political Systems of Western Malaya* (London: The Athlone Press, 1958), p. 113.
56. Liew, 'The Relationship between the Bau Kongsi and the Brooke Government', p. 172.
57. Ibid.
58. Spenser St. John, *The Life of Sir James Brooke*, pp. 63–4.
59. Keppel, *A Visit to the Indian Archipelago in H.M. Ship Maeander*, Vol. 2, p. 381.
60. Ibid., pp. 370–1. Interestingly, according to the oral traditions of the Chinese in Bau, a similar course of action was taken by the *kongsi* in February 1857 when a flotilla of boats descended upon Kuching, leading to the outbreak of war. This action resulted from a report that *a kongsi* member accused of adultery, a *kongsi* crime punishable by death, had sought refuge in Kuching. See Liew, 'The Relationship between the Bau Kongsi and the Brooke Government', p. 172.
61. Charles Brooke, *Ten Years in Sarawak*, Vol. 1, p. 27.
62. Hugh Low, *Sarawak; Its Inhabitants and Productions: Being Notes during a Residence in that Country with H.H. The Rajah Brooke* (London: Bentley, 1848;

reprinted Singapore: Oxford University Press, 1988), p. 25.

63. James Brooke to Lord Laird, 27 September 1853, MSS Pac. 590, Vol. 6, Rhodes House, Oxford.

64. For examples of this explanation, see Steven Runciman, *The White Rajahs: A History of Sarawak from 1841 to 1946* (Cambridge: Cambridge University Press, 1960), p. 126; and John Chin, op. cit., p. 34.

65. Runciman, op. cit., p. 126.

66. Spenser St. John, *Life in the Forests of the Far East*, Vol. 2, pp. 354–5.

67. *BB*, 19 June 1982.

2
Economic Changes in Bau, 1857–1898

A crucial turning point for *kongsi* mining was the *kongsi*–Brooke conflict of 1857. Details of the fighting have been recorded in many accounts.¹ A brief summary of the conflict will suffice. The *kongsi*–Brooke war had wide-ranging implications for the Chinese in general, and the Hakkas in particular. The Hakkas lost their political and economic autonomy, never to be regained, and consequently, as a dialect group, were to remain in disfavour, from the standpoint of the Brookes and other Chinese speech groups.

Most accounts² of the war emphasize the political and economic destruction of the *kongsi* after 1857. However, it will be argued here that *kongsi* mining in Bau did not decline. Hakka *kongsi* were revitalized, though under tighter scrutiny from Kuching, and on a somewhat reduced scale, stripped of the vestiges of their previous autonomy. Prior to 1857, the 'Twelve Company' had successfully retained its monopoly over gold mining, but after that year, the Hakkas faced increasing economic pressure from the Borneo Company, a major capitalist concessionaire allowed to operate in nineteenth-century Sarawak,³ and from Kuching Hokkien and Teochiu merchants. This competition slowly led to the abandonment of *kongsi* mining and a switch to cultivating the cash crops of pepper and gambier.

The Kongsi–Brooke *War of February 1857*

The contest for ultimate control of the Bau or Upper Sarawak⁴ mining district was finally resolved in February 1857. As already noted, conventional sources sympathetic to the Brookes claim that a Brooke 'tax' or fine imposed on the Bau *kongsi* for alleged smuggling of opium was the catalyst which sparked the fighting. Bau oral memories, on the other hand, state that the *kongsi* attacked the town because one of its members, wanted for adultery—a *kongsi* crime punishable by execution—was hiding in Kuching. Both versions fail to take note of the long-term differences between the *kongsi* and James Brooke.

In February 1857, a force of 600 armed miners marched down from Bau to Tundong, and from there, set out for the Brooke capital in boats. The town was attacked twice. During the first phase of the battle, the *kongsi* miners surprised the Brooke capital, meeting little resistance. They concentrated their attacks on the European—particularly the 'official'—sector of the town, avoiding the Chinese and Malay quarters. The *kongsi* wanted to emphasize that its conflict was with James Brooke. The Rajah escaped the fighting. The *kongsi* then withdrew from Kuching but met with a Malay counter-attack. This provoked the *kongsi* into retaliation, with a much bigger force this time, bolstered by additional conscripts from the gold mines and the market gardens of Sungei Tengah near Kuching. Kuching was attacked indiscriminately, and after the *kongsi* had assumed control of the town, a conference was called to apportion the responsibilities for governing Kuching and Bau, to be divided between the *kongsi* leaders, the Malay élites, and the remaining 'non-official' Europeans (traders and missionaries). However, rumour of Charles Brooke's possible retaliation with the help of his Iban warriors caused the *kongsi* to abandon its plans of governing Kuching, and to withdraw immediately to Bau. The arrival of the Borneo Company steamer, *Sir James Brooke*, together with the predicted appearance of Charles Brooke and his Iban allies, saved the Brooke capital. The *kongsi* was put to flight and beat a hasty retreat to Bau and over the watershed into Sambas. Considerable loss of life was suffered by the *kongsi*, and it was estimated that some 3,500 men, women, and children perished at the hands of the Ibans or were driven off into Kalimantan.[5]

Revitalized Kongsi *Mining*

The collapse of Bau as an autonomous Hakka mining district corresponded directly with the economic ascendancy of Kuching. Indeed, Kuching commercial interests were to benefit most from the enlarged economic opportunities, previously the monopoly of the Hakka mining *kongsi*. The Borneo Company, with its superior resources and the Rajah's backing, was able to swamp erstwhile *kongsi* interests. The economic position of Kuching's Hokkien and Teochiu merchants was strengthened, with new opportunities being presented in the cultivation of pepper and gambier, and in the role of Kuching as an entrepôt for an expanded hinterland embracing Bau. The Kuching traders, with their expanded resources and taking advantage of trading opportunities elsewhere, were able to expand their mercantile activities in a diaspora that covered other riverine basins of Sarawak. The Hakkas were to remain a mining and farming community.

In the aftermath of the war of February 1857, the mining district lay in ruins, but it was not long before the mines and gardens were re-established. Hakkas from Lundu, Sadong, Kuching, and from across the watershed returned to the abandoned mines. The leaders of the reconstituted *kongsi* had no choice but to establish an amicable relationship with the Rajah, as Bau now came directly under the political control of Kuching.[6]

Kuching merchants provided financial credit to the revitalized mines and gardens. The extending of credit created a patron–client relationship between the financiers and the miners and market gardeners, with certain social obligations and loyalty to the creditors ensured. One of the Kuching leaders of the post-1857 period, a Chao-ann by the name of Chan Ah Koh, was a trader with considerable business interests in Bau.[7] Ghee Soon and Company, a major Kuching firm owned by two Teochius, Law Kian Huat and Sim Ah Nio, extended its sphere of operations into Bau, being principally involved in financing gambier and pepper gardens.[8] There were other minor Kuching interests, as reports from the *Sarawak Gazette* indicate.[9] It would appear that the Kuching traders were mainly involved in financing gambier and pepper gardens, as cash-cropping slowly began to replace mining as a source of livelihood.

An important credit supplier, as will be seen, was the Borneo Company. From the outset, it was authorized 'to work mines, ores, veins, or seams of all descriptions of mineral in the island of Borneo and to barter or sell the produce of such working'.[10] In later years, the company enjoyed other concessions and trading privileges in activities as diverse as banking, shipping, and sago, indigo, and tobacco cultivation. In March 1879, the company was given the monopoly of using quartz machinery and prospecting for gold for a period of fifteen years.[11]

It is not known how active in mining the *kongsi* were between 1857 and 1870,[12] but from the latter period onwards, sufficient evidence exists to suggest that the *kongsi* were thriving. A government census of 1870 indicated a revitalized Hakka presence in the Bau district. A total of 1,145 Chinese was counted in the gold, antimony, and mercury mines, spreading from Bau to Buso, Jambusan, Paku, and Tegora.[13] Kuching, in comparison, had a Chinese population of 2,251 counted in 1876.[14] However, the majority of the Chinese in Bau were employed by the Borneo Company. At Jambusan, Paku, and Buso, the company had no fewer than 750 men, with another 200 working for contractors, making a total of 950.[15] It can be surmised that the rest of the Chinese unaccounted for by the Borneo Company worked on their own, or in *kongsi*. However, even *kongsi* operating on their own were unable to compete with the Borneo Company, which was better endowed with capital and equipment. In 1870, a *kongsi* at Piat was reported to have been unsuccessful in its mining activities, and had ceased working entirely.[16] A *kongsi* called the Soon Hen *kongsi* stopped mining altogether and decided to work on a contract basis for the Borneo Company in the same year.[17] Bau once again took on a distinctive Hakka demographic appearance with some 200 Chinese houses counted in 1872.[18] Indeed, the Hakkas after the war of 1857 were as active as the pre-1857 miners in physically transforming the hills of Bau by building reservoirs, damming up valleys, and constructing devices such as locks and floodgates.

By 1876, a clearer picture of Hakka resourcefulness in mining is obtained. In Paku alone, there were four *kongsi*. These were not entirely Chinese in composition, but had Malay members as well.[19] Observers described them as co-operative ventures, pooling their resources and sharing their profits:

> The kongsi have at different times made considerable profits, but their system of paying away everything earned during the year as divided at their yearly wind-up, as well as the custom that commonly obtains of advancing cash to the shareholders in anticipation of profits frequently lands them in acute difficulties. Another custom is that if during the working of a kongsi, a shareholder parts with a portion of his shares to another person, the original shareholder, not the purchaser, is considered responsible for any loss.[20]

The organizational structure of a typical *kongsi* consisted of a headman, a clerk, a cook, a carpenter, and a tailor, in addition to the mine labourers. It was noted that 'they divide the gold every three or four months, according to agreement, the debts for the advance of money and provisions being first deducted'.[21] The post-1857 *kongsi* still retained their frontier characteristics of partnership and brotherhood, but they did not exercise any political autonomy, nor were they allowed to this time.

Around the period of the mid-1870s, many small *kongsi* were in existence. In 1876, at Piat, the two biggest *kongsi* had 47 and 42 members respectively. In addition, there were five other smaller *kongsi* with a total of 113 labourers. During the same year, in Paku, there were two *kongsi*, one with 40 members, and the other with 22. There were 40 other individual miners on their own.[22]

Bau had two rival companies at work, the *kongsi* Si Parit Besar (Tai Parit *kongsi*) and Shak Luk Mun (Golden Dragon Gate) *kongsi*. It is not known what the sizes of these two *kongsi* were, although it can be conjectured that each might have had several hundred members. These two *kongsi* had separate villages and shops, with the latter described as 'well-built and neatly kept'.[23] Although there were many *kongsi* in the 1870s, there was only one big *kongsi*, the 'Twelve Company', before 1857. In 1848, the 'Twelve Company' had about 400 persons under its employment,[24] and was far larger than any of the *kongsi* after the 1857 war. That there was only one major *kongsi* in the pre-1857 days could have been due to its superior organizational ability and autonomy. The formation of multiple *kongsi* after 1857 was due to the loss of autonomy of the Hakkas, the district now being subject to outside pressure and competition, as well as tight political control by the Rajah's men.

The Brooke Presence in Bau

With the *kongsi* and its leadership destroyed just after the war of 1857, James Brooke assumed direct control of the Bau district for the first time. A fort was constructed at Belidah, and a European officer administered the district until 1861 when a native police force was subsequently put in charge. In 1871, the fort was dismantled and shifted to Paku. Paku served as the administrative centre of the district until 1898, when Bau took over.

The fort at Paku overlooked the scattered Hakka settlements, and was manned by six to eight policemen. An officer-in-charge, with policing and magisterial functions, was responsible for looking after the Hakka settlements of Jambusan, Bau, Buso, Bedi, and Tegora, the last with a mixed populace of Hakkas, Malays, and Land Dayaks who were employed by the Borneo Company in excavating cinnabar. The Brooke officer-in-charge of Paku, Noel Denison, in a tour of the district in 1886, said that the Hakka gold and antimony workers were 'a somewhat difficult people to rule', but nevertheless praised them by saying: 'It is pleasant to see them so active and hardworking, their praiseworthy diligence being plainly seen in their numerous mines, gardens and clearings; when working for themselves, they are no doubt, the most pushing and industrious race of Asiatics.'[25] The post-1857 Bau Hakkas were performing a useful pioneering role in the opening up of the district.

While some *kongsi* leaders co-operated with the Brooke regime, there were others who did not. Organizations considered a threat to the regime were classified as secret societies. In 1870, Charles Brooke passed an ordinance making it a capital offence to be a leader of a secret society.[26] Gin Vong, head of an organization deemed to be a secret society, was sentenced and banished from the country in 1869; he was rearrested and executed in April 1876.[27] A 'secret organisation' was broken up in July 1889 at Panto Pajang and S'gobang.[28] This society was located in the centre of a planting district, and most members of the group were said to be from Mandor, with which they had 'regular communication'.[29] The organization reportedly had hundreds of members organized and ready for service. The organization must have been considered a serious threat to Rajah Brooke, for six of its leaders were executed and eleven others were sentenced to life imprisonment.[30] By the end of 1906, four 'secret societies' had been discovered and their leaders executed; the societies were then disbanded. The memory of the 1857 conflict was still fresh in the Rajah's mind, and Charles Brooke was wary of any organization which appeared to be run like the politically independent *kongsi* of the pre-1857 days.

The political decline of the Bau Hakkas was matched by a corresponding drop in their economic control of the district.

The Hakkas and the Borneo Company

The Borneo Company moved into the Bau district immediately after the 1857 conflict. Initially, the company was principally involved in antimony and cinnabar mining.[31] In its formative years, the company was not involved in gold mining, but instead made regular purchases of the mineral ore from the local miners in Bau.[32] Initially, too, the Borneo Company was at pains to avoid direct conflict with *kongsi* interests, and was not willing to openly wield its economic power. It judiciously financed and supervised separate 'tributor' *kongsi* around Buso. Each *kongsi* was paid fixed rates in return for the ore delivered to the company.[33]

As the concessionaire's investment interest in gold grew, it ran into conflict with the Hakka miners. Relations between the Borneo Company and the largest *kongsi* in the district, the Shak Luk Mun *kongsi*, became difficult as the Hakkas resented and objected to the advancement of the company into Bau, 'and ore stealing, labour crimping and finally interference with the Company's water supply hampered the Company considerably'.[34] Around 1884,[35] the Borneo Company bought out the *kongsi*, took over its existing workings, and secured a monopoly of all gold workings in the state. Even before it did that, it already had an advantage over the existing *kongsi* with its superior mining technology, and was in the position of being able to hire out equipment to the Hakka miners. For example, in 1884, the Choo-Tiong *kongsi* at Piat hired an engine of 8 hp from the Company—it was probably used for pumping water—while at

Paku, the Hian Chong *kongsi* hired a 30 hp engine.[36]

The European-owned enterprise was not to achieve full control of the gold mines until 1898, when it introduced an innovative method of extracting gold ore by the cyanide process. Preparatory groundwork for the cyanide machinery, at the site formerly occupied by the Shak Luk Mun *kongsi*, started in 1896. The company, in that same year, moved in to buy out the reservoir and works of the Shak Luk Mun *kongsi*'s rival, the Tai Parit *kongsi*.[37] By 1897, it had already bought over the remaining principal *kongsi* at Jambusan—the Shoon Fat, Shoon Hin, and T'hap Shak *kongsi*—as well as acquiring a considerable piece of land near by.[38] By the end of 1898, the first cyanide factory was completed.[39] A second plant was built at Bedi in 1900.[40] By the time the Borneo Company closed down its Tai Parit open-cast mine in 1923, the mine had yielded about half a million ounces of gold ore.[41] Besides mining activities, the Borneo Company had expanded its sphere of operations to include agricultural activities as well. Its superior technological resources and financial prowess brought about an eclipse of the *kongsi* by the turn of the century. An important factor which helped the company achieve economic control of the Bau district was the backing of the Brookes.[42] But even without the support of the Rajah, the company was almost certain to have its own way by virtue of its economic power.

The Shift to Crop Cultivation

The expanding, monopolistic practices of the Borneo Company were to force the Hakka gold miners to turn to pepper and gambier planting. An additional pressure was that the population of the Bau district was expanding. This demographic change was also responsible for the structural economic shift to cash-cropping. Pepper had long been a traditional crop in Borneo and was reported to have been grown almost a century before in Brunei.[43] In Sarawak, in the area now covered by the present Kuching and Samarahan Divisions, pepper became an attractive cash crop from the 1870s onwards, when Charles Brooke promulgated a series of liberal land laws and planting regulations between 1872 and 1876,[44] and encouraged Kuching's leading *towkay* to pioneer and finance the industry. The first areas to be cultivated were Lundu and the outskirts of Kuching. The pioneer planters in these areas were Teochius. It would appear that pepper and gambier planting in the Hakka-populated district of Bau was only taken up in the 1880s.

One of the many Teochiu and Hakka pioneers who took the initiative in cultivation was a Hoppo Hakka, Pui Shin Wen, who arrived in Bau in 1891. Pui initially lived in Siniawan, and worked as a coolie in a pepper garden. Not being conditioned to the hot and humid conditions of the tropics, he recounted experiences—perhaps somewhat exaggerated—of having to take two baths a day, of one hour's duration each. He received wages of 40 cents a day. The following year, 1892, he left for Bau to work in the gold mines.[45]

A typical example of a *sinkheh* who died poverty-stricken—and there must have been many others like him—was another Hoppo Hakka, Lee Sui, who migrated to Sarawak in 1877, at a time when Charles Brooke was inviting Chinese pioneers to take up planting. He first journeyed to Kuching under terrible conditions. Fresh water had to be stored in bamboo pipes and, when it ran out, the boat passengers had to use sea water. Unlike most other Hakkas who gravitated towards Bau, Lee made his way up to Betong, where he unsuccessfully tried growing padi and then switched to pepper. The story of his later life was a familiar one for many sojourners from China who never fulfilled their dreams of amassing wealth in Sarawak. The *sinkheh* never married due to the pressure of hard work, failure to save up sufficient money, and an addiction to opium. He spent his remaining days at an old people's home in Kuching.[46]

A Hakka sojourner who recounted hardships of life in the *Nanyang* to his kindred in his homeland and did not paint a rosy picture as many were prone to do, was Lee Kai Ku, born in 1877; he arrived in Bau in 1900, and was reputed to be a hard-working man, capable of tackling two to three jobs a day. Lee told his kinsmen who intended to go abroad that 'it was not easy to earn money in the *Nanyang*, you had to sweat a lot; during the daytime, your clothes were never dry'.[47]

These oral accounts, besides narrating pioneering experiences, reveal the extent of economic and demographic changes affecting Bau in the second half of the nineteenth century. A new source of settlers had come directly from China.[48] The Hakkas had a reputation for industry and endurance, qualities which put them in good stead for the tough pioneering work of opening up new land. Many Hakkas had taken up planting. While the newcomers from China took to planting, those from Kalimantan had originally been miners, forced to switch to cultivating partly because of the monopolistic practices of the Borneo Company, and partly in response to high prices for cash crops.

The interest in growing pepper could be seen from the increasing number of applications for permits for land titles. In the single month of May 1886, there were fifty applications for permits to plant pepper.[49] Even then, many pepper and gambier gardens remained unregistered. It was stated in the *Sarawak Gazette*[50] that many of the gold-miners were turning their attention to pepper. These pioneer gardens were moderate in size. To take an example, three miners, Bau Soon, Ah Wong, and Ah Look, who each planted pepper in the neighbourhood of Tegora in August 1896, had 4,000 plants in the first two instances, and 1,000 in the third.[51] It was reported that 'the extension of pepper gardens all over the country was now the order of the day'.[52]

Not all gardens were established through individual initiative and expense. As already noted, Kuching merchants provided financial credit for the gardens. In 1889, Reginald Awdry, Resident of Upper Sarawak (Bau), discovered six pepper gardens near Sungei Kandis, opposite the mouth of the Siniawan River. These gardens, as well as a number of others in the district, were financed by Chin Ann and Company of Kuching.[53] Indicative of the interest in pepper towards the end of the nineteenth century was the large number of legal suits brought by planters before the courts, both in Kuching and Bau, to restrain others from opening up new gardens or from encroaching on existing gardens by felling old jungle in the vicinity.[54]

The shift to pepper planting had its repercussions on the mining industry, and on Chinese relations with the Land Dayaks. As pepper vines were grown indiscriminately without permits from the understaffed government offices, Charles Brooke issued an order in 1887 to residents of Upper Sarawak, encompassing the areas of Buso, Paku, Bau, Bedi, Jambusan, and Tegora: 'Anyone found planting pepper in the above-named districts does so at his own risk, and in the event of the land being required for mining purposes either directly or indirectly, he will not be entitled to any compensation.'[55]

The Hakka gardeners faced a dilemma. Besides being excluded from mining interests which were fast becoming a monopoly of the Borneo Company, Land Dayak social mores on land tillage and ownership posed an acute problem for the prospective planters. Land Dayak land rights were established by persons who first cleared the primary jungle and these rights were subsequently passed down to the descendants.[56] Similar land disputes had been reported in 1887 in Lundu, in another area of the First Division. It was reported in the *Sarawak Gazette*: 'Land disputes and collision between Chinese gardeners and Dayaks are somewhat constant, and from the number of pepper gardens opening all round Lundu, such disputes are bound to increase.'[57]

The land disputes between the Hakka planters and the Land Dayak swidden agriculturists were serious enough for Charles Brooke to make a personal visit to Bau to arbitrate on the matter in 1896:

He sat in court and spoke at length to the assembled Dayaks and Chinese. The former had specially come to complain of the encroachment of the Chinese pepper planters upon lands hitherto cultivated by them once in every 12 to 15 years, and which at best produced a precarious crop of rice at the cost of the total destruction of all timber and other jungle stuff produced on the land. The Rajah directed that the Chinese pepper planters should be encouraged.[58]

Charles Brooke, who wished to promote agricultural growth in the state and looked to the Chinese as the precursors of economic development, was not prepared to admonish the Hakka gardeners, even if it meant that Land Dayak customary law was being violated. The basic causes of the land disputes were the Chinese lack of understanding of Land Dayak cultural mores, and the increasing needs of the planters to open up land in areas which outwardly looked abundant and empty, but in fact were claimed by the Land Dayaks through their customary law.

Besides gaining a monopoly on mining, the Borneo Company took an early interest in establishing control over the pepper crop. It acquired suitable land concessions in the vicinity of Buso. Individual plots were leased out to Chinese and Malay gardeners and, by a discreet policy of judicious advances,[59] the company ensured for itself a steady and constant supply of the peppercorns. Company staff at Buso made constant visits to the gardens to supervise the cultivation:

Not wishing to advertise their growing monopoly unnecessarily, the Company's employees were encouraged to take out loans from the Company bearing interest at 5% per annum. This enabled them personally to take up mortgage from the owners of the local gardens on a return of 12% per annum. The Company made 5% per annum on the capital advanced and secured some measure of control over the crops, while the assistant was permitted to retain 7% as personal remuneration.[60]

When there was a sudden increased demand in the 1890s for gambier, which was used in tanning, dyeing, and as food seasoning by the Chinese, the company secured a 20,000-acre concession at Poak. Gambier was a cash crop which complemented pepper well. Old gambier plants, when burned, could be used as fertilizer for pepper. It was estimated by the company that a pepper garden of 1.5 acres needed 50 acres of gambier. With this expansion of Borneo Company interests, a labour force had to be recruited, and this came through a regular system of *sinkheh* obtained through Singapore. T. C. Martine, an ex-manager of the Borneo Company, recollected:

A man was brought over and guaranteed employment, for a period of say, three years, during which time he was legally bound to the Company's service and was liable to imprisonment for absconding. Reviewing the matter over a long period of years, this type of mild slave labour worked well and discontented coolies were not numerous, or at anytime troublesome.[61]

It was not true that discontented labourers did not protest against the injustices of *sinkheh* bondage. In 1907, a group of over 100 Liuchius working for the Borneo Company were expelled for being a 'truculent lot'.[62]

Besides the Borneo Company, there were many smallholders who were still able to hold out on their own. In 1900, a total number of 1,049 gardens consisting of 908,230 vines was counted.[63] Seven years later, another census accounted for 1,320 gardens. However, by the beginning of the twentieth century, structural shifts in the general economy of the state due to the popularity of rubber were being felt in Bau, and pepper gardens were then abandoned. In 1907, in the Bau district, it was reported that 187 gardens had gone out of cultivation and 'numbers of gardens have suffered from neglect and will produce smaller crops ... at least 250 gardens will be abandoned by the end of the year'.[64]

By the turn of the century, the drastic occupational shift for the Hakkas, from mining to cash-cropping, had taken place. Many Hakkas, unable to compete with the Borneo Company, had sold off or abandoned their mines and became smallholders of gambier and pepper farms. The Borneo Company had gained control of mining and cash-cropping in the Bau district.

1. The accounts, invariably pro-Brooke, from the writings of observers, Brooke officials, historians, and local writers, come from the following: Spenser St. John, *Life in the Forests of the Far East*, Vol. 2 (London: Smith Elder and Co., 1862; reprinted Singapore: Oxford University Press, 1986), pp. 336–64; Ludvig Verner Helms, *Pioneering in the Far East, and Journeys to California in 1849, and to the White Sea in 1878* (London: W. H. Allen, 1882), pp. 164–92; Harriette McDougall, *Sketches of Our Life in Sarawak* (London: Society for Promoting Christian Knowledge, 1882), pp. 125–6; Steven Runciman, *The White Rajahs: A History of Sarawak from 1841 to 1946* (Cambridge: Cambridge University Press, 1960), pp. 125–33; and John Chin, *The Sarawak Chinese* (Kuala Lumpur: Oxford University Press, 1981), pp. 27–34. The conflict is conventionally described as the Chinese Rebellion of 1857.

2. See note 1.

3. The role of the Borneo Company in the economic history of Sarawak has been much neglected. As an *imperium in imperio* after its formation, it had much economic influence in Kuching and Bau which the Brookes must have found difficult to counter. A shrewd observer in 1878 noted:

'The original and benevolent idea of Sir James Brooke, that the Dayaks could be civilized and induced to develop the resources of the country is almost abandoned, and the foolish chimera of the sort of political sovereignty is steadily giving way before the business-like pursuits of the commercial gentlemen [of the Borneo Company] who are now the practical master of the country.'

(Anonymous, 'The Chinese in Borneo', *The China Review*, Vol. 7 (1878–9), p. 8.)

4. Following official usage, it means the Upper Sarawak River, which meanders through and drains the Bau district.

5. Spenser St. John, *Life in the Forests of the Far East*, Vol. 2, p. 358.

6. *SG*, 16 March 1913, p. 159; and 1 November 1926, p. 283.

7. *SG*, 31 March 1959, pp. 51–2.

8. *SG*, 1 May 1897, p. 95; and Craig Lockard, 'The Southeast Asian Town in Historical Perspective: A Social History of Kuching, Malaysia, 1820–1970' (Ph.D. thesis, University of Wisconsin, 1973), p. 201.

9. *SG*, 1 May 1896, p. 95.

10. Cited in Sir Percival Griffiths, *A History of the Inchcape Group* (London: 1977), pp. 133–4.

11. H.H. The Rajah's Order Book, 1863–1890, p. 211, Sarawak Archives (SA).

12. There is no mention of *kongsi* in the Brooke records during this period between 1857 and 1870. The government publication, the *Sarawak Gazette*, started publication in 1870 and only after that date did reports on the *kongsi* appear.

13. Monthly Reports, 1 October 1871, pp. 331–2, SA.

14. H. Wilfred Walker, *Wanderings among South Sea Savages in Borneo and the Philippines* (London: Witherby, 1910), p. 188.

15. Monthly Reports, 1 October 1871, p. 331, SA.

16. Monthly Reports, 1 October 1870, p. 105, SA.

17. Ibid.

18. *SG*, 31 January 1872, p. 12.

19. *SG*, 8 September 1876, p. 14.

20. *SG*, 1 April 1897, pp. 70–1.

21. *SG*, 1 June 1886, p. 98.

22. *SG*, 17 July 1876, p. 3; and 8 September 1876, p. 14.

23. William Hornaday, *Two Years in the Jungle* (London: K. Paul, Trench, 1885), p. 479.

24. Spenser St. John, *Life in the Forests of the Far East*, Vol. 2, p. 323.

25. *SG*, 1 June 1886, p. 108.

26. Sarawak Government Orders, 1860–1891, 14 May 1870, p. 43, SA.

27. *SG*, 22 April 1876, p. 1.

28. *SG*, 1 August 1889, p. 108.

29. S. Baring-Gould and C. A. Bampfylde, *A History of Sarawak under Its Two White Rajahs 1839–1908* (London: Sotheran, 1909; reprinted Singapore: Oxford University Press, 1989), pp. 204–5. In 1884, the Lan-fang *kongsi* fought a war with the Dutch in which the *kongsi* lost. It is possible that some of its members crossed the watershed into Bau.

30. Secret societies appear to have been formed among the Hakkas, Foochows, Liuchius, and Cantonese, dialect groups which were dominant among the rural planters and the urban working class. See *SG*, 22 April 1879, p. 1.
31. *SG*, 16 January 1877, p. 7.
32. T. C. Martine, 'History of the Borneo Company Ltd', typescript, Borneo Company Limited, London. From 'Notes written by me when in Singapore (Changi Gaol) 1943-4', p. 21.
33. Ibid.
34. Ibid.
35. Henry Longhurst, *The Borneo Story: The History of the First 100 Years of Trading in the Far East by the Borneo Company Limited* (London: Newman Neame, 1956), p. 67.
36. *SG*, 1 May 1884, p. 39.
37. *SG*, 2 March 1896, p. 80; and 1 June 1896, p. 117.
38. *SG*, 1 April 1897, p. 70.
39. *SG*, 1 December 1898, p. 195.
40. *SG*, 1 May 1900, p. 83.
41. *SG*, 31 October 1956, p. 245.
42. Letters of Rajah Charles Brooke, Vol. 5, 15 October 1900, pp. 278-9, SA.
43. Thomas Forrest, *A Voyage to New Guinea and the Moluccas from Balambangan, including an Account of Magindano, Sooloo and Other Islands* (London: G. Scott, 1779), p. 40.
44. *SG*, 13 June 1872; and H.H. The Rajah's Order Book, 1863-1890, pp. 120-4, SA.
45. Liew Peck Kwee, *A History of the Hoppo Chinese with Special Reference to Sarawak* (Singapore: Tung Yet Publishing House, 1978), pp. 75-6 (text in Chinese).
46. Ibid., p. 77.
47. Ibid., p. 79.
48. In 1900, Charles Brooke made arrangements with a Sin-Onn Hakka named Kong Kui En to introduce some 300 Sin-Onn families as pioneering planters to settle in the suburbs of Kuching. See *SG*, 31 March 1959, p. 51.
49. *SG*, 1 May 1886, p. 72.
50. *SG*, 1 January 1889, p. 2.
51. *SG*, 1 August 1896, p. 143.
52. Ibid.
53. *SG*, 1 November 1889, p. 102.
54. *SG*, 1 November 1895, p. 193.
55. *SG*, 1 September 1887, p. 144.
56. W. R. Geddes, *The Land Dayaks of Sarawak: A Report on a Social Economic Survey of the Land Dayaks of Sarawak* (London: HMSO, 1954), p. 59.
57. *SG*, 1 July 1887, p. 121.
58. *SG*, 1 June 1896, p. 117.
59. Martine, op. cit., pp. 37-8.
60. Ibid.
61. Ibid., p. 5.
62. *SG*, 4 March 1907, p. 54.
63. G. Dalton, 'Pepper Growing in Upper Sarawak', *SMJ*, Vol. 1, No. 2 (February 1912), p. 54.
64. *SG*, 5 June 1907, p. 141.

3
Traditional Patterns of Trade in Borneo

THE island location of Borneo attracted Chinese, European, and regional traders while the local trading networks were monopolized by the Malays. The stimulus to trade was the abundant natural resources of the island, and a local population receptive to goods coming in from the outside.

The Historical Patterns of Chinese Trade with Borneo

Evidence of Chinese traders conducting their mercantile activity with countries south of China stretches as far back as the first century BC.[1] South-East Asia was ideally located at the crossroads of the world trade routes between the Middle East and India, and China. South-East Asian ports and harbours, taking advantage of their strategic locations, emerged to serve as entrepôts and international trading centres for the trade routes between East and West. Archipelago island kingdoms, predicated upon commerce, sprang up. In Palembang, South-east Sumatra, Srivijaya (seventh to twelfth centuries) rose as a dominant trade and political empire, followed by Majapahit (thirteenth to fifteenth centuries) in Central Java, and Malacca (fifteenth to sixteenth centuries) on the west coast of Malaya. Archaeological evidence in Sarawak suggests that the north-western and northern coast of Borneo could have participated in the main currents of South-East Asian trade as early as the eighth century. Chinese sources can also help to identify place-names in Borneo which had contacts with the Celestial Empire of the Ming Dynasty (1368–1643) within the framework of 'tributary' relations. Lastly, seventeenth- and eighteenth-century European written accounts will be relied upon for their descriptions of a contemporary Chinese presence in Borneo.

Extensive archaeological diggings have been carried out at the Sarawak River delta and the Niah Caves since 1947. Minor finds have been recorded in the coastal area of Kabong, and inland at Gedong, 35 miles from the mouth of the Sadong River, and further up-river at Bukit Sandong and Ensika.[2] In the Niah Caves in northern Sarawak, several burial sites containing seventh- to eighth-century Chinese ceramic export-ware have been discovered.[3] (The Niah limestone

Map 4 Archaeological Sites in Sarawak

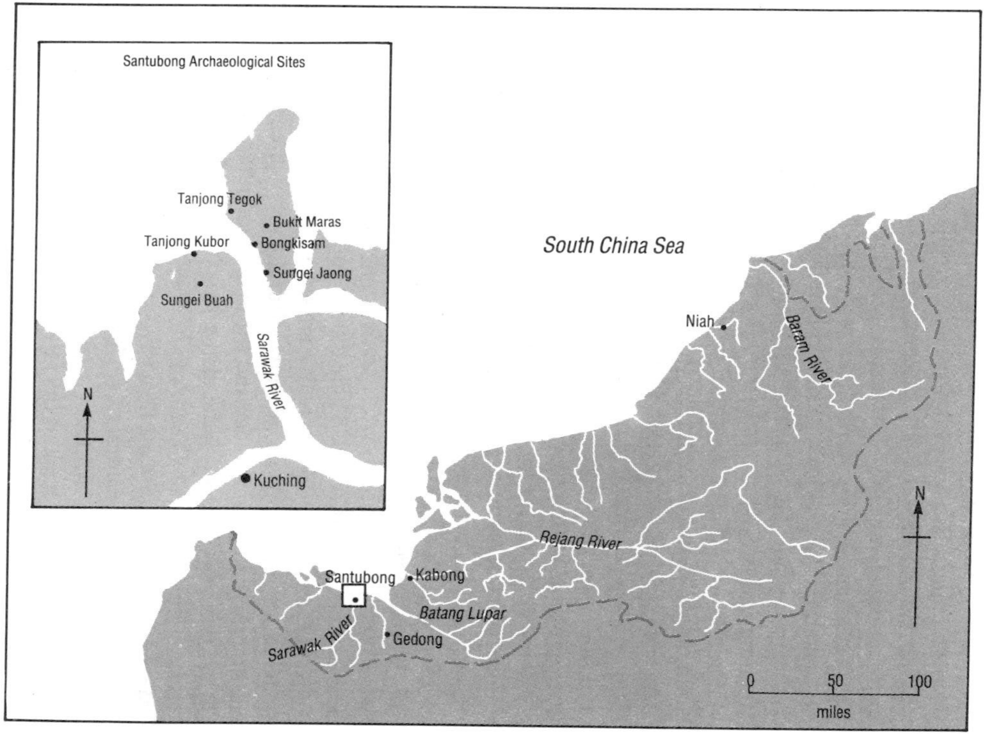

cliffs still serve as home to thousands of swiftlets whose nests, moulded from an edible salivary secretion, are highly valued in Chinese cuisine.)

The most significant archaeological discovery in terms of revealing patterns of Chinese trading contact with Borneo has been made at Santubong, at the mouth of the Sarawak River, 18 miles down-river from Kuching. The Sarawak River delta, with its access to the sea, was in an ideal location as a trade centre and a victualling site for ships awaiting the change of monsoons. The hinterland in Bau was rich in bird's nest and gold, commodities in demand in China. Thousands of broken ceramic pieces, beads, coins, gold and silver ornaments, and large quantities of iron-slag, remnants of an erstwhile iron-smelting industry, have been found at various sites in the delta— at Santubong, Sungei Jaong, Bongkissam, and Bukit Maras. The abundant material evidence on the ground suggests that the traders who came to the river-mouth could have been Chinese and Indians.[4] The excavated ceramic pieces consist mainly of glazed earthenware and stoneware jars, plates, dishes and bowls, dating back to the Tang (618–907) and Sung (960–1297) Dynasties.

The archaeological finds of ceramics and iron-slag indicate that an industrial-cum-trading zone existed at the mouth of the Sarawak River, with its peak period possibly in the twelfth century, as evidenced by a study of the excavated ceramics, coins, and ornaments, and by radio-carbon dating of these items.[5] The iron ore and ceramics were sent inland and up-river, traded for jungle products and other exotic products of Borneo. There is little doubt that this trade existed. That old ceramics are still kept by Sarawak's indigenous communities, playing an important social and cultural role in their lives, bears testimony to the significance and the long-standing impact of a trade in ceramics in Borneo. Some of the jars still intact in their entirety and highly valued by Sarawak's interior communities are the Martaban jars[6] of the Tang Dynasty, coarse earthenware jars covered with a brown or green glaze. A 1967 field-study of ceramics kept by Melanaus on the Mukah coast revealed that most of the ceramic pieces can be dated back to the fifteenth century or later, and consist mainly of Ming and Ching celadons in the form of dishes, lidded boxes, and jarlets, and a few jars of Siamese (fourteenth-century Martaban imitations), Japanese, and European manufacture.[7]

Archaeological evidence from elsewhere in Sarawak, apart from that found in Santubong and Niah, supports the idea of a lineal continuation in the ceramics trade. From 1967 to 1976, archaeological field-work was undertaken at fifteen new sites, resulting in further discoveries. Annamese monochrome wares of thirteenth- and fourteenth-century origin have been found at Tebing Tinggi, Kabong; Siamese Sawankhalok wares (dating from the fourteenth and fifteenth centuries) at Bukit Sandong, Gedong; and Chinese and Annamese blue-and-white wares (from the sixteenth to nineteenth centuries) at Bukit Sandong and Murah in Kabong.[8] These findings are important for they strongly suggest a continuity in Chinese trading activity, stretching along the coast of the present-day Kuching and Samarahan Divisions to parts of the Sri Aman Division. Chinese historical sources can be referred to for documented evidence of officially sponsored contacts between China and other parts of northern Borneo under the system of 'tributary' relations. It is believed that a state by the name of P'o-ni, possibly located at the site of present-day Brunei, had 'tributary' relations with the Celestial Empire.[9]

P'o-ni was first mentioned by Chinese sources in the tenth century. In 977, the Sultan of P'o-ni sent envoys to the imperial court of Peking, as a result of a merchant (presumably Arab) having visited P'o-ni *en route* from China. The Sultan sent tribute, expressing a desire to continue the practice, and asked that the Emperor of China protect P'o-ni's ships passing near Champa (Indo-China). In 1082, the Sultan again sent tribute to China.[10] There is no record of official contact between P'o-ni and China during the Yuan Dynasty (1280–1367), although a disastrous military expedition was sent by the Chinese to Java in 1292. With the well-known voyages of the eunuch admiral, Cheng Ho, the emperors of the Ming Dynasty expanded their contacts with the *Nanyang*. In 1408, the Sultan of P'o-ni visited China and had an audience with the Emperor of China. The Sultan died before the homeward journey. His son, Hiawang, returned to P'o-ni with his mother, but later paid another visit and continued to send tribute.[11]

Chau Ju-Kua, a Ming court chronicler, has given descriptions of P'o-ni based on information from travellers who had visited the place. P'o-ni possessed products like camphor, wax and tortoiseshell, which were bartered for Chinese gold, silver, silk, beads, and ceramics. Chau described how when a Chinese junk arrived in P'o-ni, the Sultan and his court were welcomed aboard and lavished with presents. Whenever the ship's crew went on shore, it was customary for them to offer the Sultan gifts of food and liquor. After about a month, the ship's crew would request the Sultan and his court to fix the prices of their trade goods. Once this had been done, drums were beaten in order to announce that permission was granted to trade on board the ship.[12]

Chinese references to P'o-ni are significant in that they constitute evidence of the existence of a strong local centre which had trade relations with China. While Santubong declined after the twelfth century—a tentative conclusion based on the failure to find any ceramics and artefacts after that period—P'o-ni continued to exist during the Ming period. At Kota Batu, 1½ miles east of Brunei's present capital, Bandar Seri Begawan, archaeological finds of Chinese coins of the Sung period[13] and pieces of Tang[14] and Ming pottery indicate a continuity in Chinese contacts with this northern part of Borneo from the ninth to the fifteenth centuries.

By the sixteenth century, when European traders had begun to participate in the dynamics of trade in Borneo, the Chinese were considered serious rivals. European traders have left useful accounts of their observations of the early Chinese presence in Borneo.

Pigafetta, writing on the landing of Magellan's fleet in Brunei in 1521, commented on Chinese influence in Borneo, particularly in Brunei, remarking on Chinese silk, Chinese weights and measures, and coins stamped with Chinese characters.[15] In 1600, Van Noort, one of the earliest Dutch traders to set foot in Borneo, spoke of Dutch failure to capture the Borneo cloth market because 'the Borneans were fond of China linen, but that which came out of Holland was a mere dreg'.[16] The inclusion of Chinese cloth as a trade item by both Pigafetta and Van Noort suggests that it was another important commodity, in addition to ceramics.

An expanded trade list was given by a Captain Beeckman when he sailed to the port of Banjermassin in southern Borneo in 1713, on a mission of making that port an intermediate station for the trade with China. Although they concern another area of Borneo, Beeckman's remarks give details of a strong Chinese presence. He wrote: 'the Chinese are the only persons that have shops tolerably well furnished; they set them off with coarse chintz, baftees, tea, drugs, china ware and many other things'.[17] The products coming out of Banjermassin were pepper, bezoar stones, gold, rattan, camphor, and diamonds.[18]

Another observer, Roggewein, continued in 1721 to talk of the Chinese dominance of trade in Borneo, adding that, ' ... they carry on a great commerce, and furnish the inhabitants in return with silks, chintz, calicoes, in short, all the manufactures of Japan and China'.[19]

The period 1690–1730 was the heyday of Chinese junk trade with Batavia, and it might not be inconceivable that some of these junks could have stopped *en route* in south-eastern Borneo, attracted to the products and trade of the area.[20] In fact, the Dutch East India Company faced considerable difficulty in attempting to gain a monopoly of the junk trade of Batavia, as the junks went to places as diverse as Ambon, Banda, Celebes, Borneo, Sumatra, and the Malay peninsula.[21]

Travellers like Forrest commented on the 'considerable commerce between China and Borneo' in the second half of the eighteenth century, with junks carrying back to Amoy quantities of black timber, wood, rattan, *damar*, resin, cloves, tortoiseshell, and bird's nest.[22] Before the end of the eighteenth century, it was estimated that five to seven junks of 500–600 tons burden sailed annually from Amoy to Brunei. By 1809, however, no junks were reported to have visited Brunei for a number of years.[23] With the rise of Sulu as a regional trading zone by 1800,[24] the traffic in trading junks to Brunei declined. The Sulu archipelago, by the end of 1800, was visited by one to four junks annually, and by 1830, the number was two junks of up to 800 tons weight.[25] By then, with the emergence of the *kongsi* on the west coast of Borneo, trade was not as important an economic activity as gold mining.

Trade with the interior natives of Sarawak became an important Chinese pioneering activity when the modern Brooke state was established in 1841. However, it was in fact the continuation of a centuries-old pattern of economic exchange between Chinese traders and indigenes. From the Chinese, the natives obtained ceramics, luxury items, and foodstuffs. Chinese traders came to collect the assorted exotic jungle products of Borneo, such as bird's nest, bezoar stones, rhinoceros horns, *damar*, and rattan.

But the Chinese were not alone in pioneering local and regional trade: the Malays, too, played a crucial role.

Traditions of Inter-island, Coastal, and Riverine Trade in Borneo

The preceding section dealt with the presence of Chinese trade in Borneo before the nineteenth century. This section concentrates on indigenous inter-island, coastal, and riverine trade in Borneo. It is necessary to establish the background for these two types of trading—that is, Chinese and indigenous trading—first, in order to show how they overlapped in the nineteenth century, and secondly, to demonstrate how Chinese traders established themselves all over Sarawak, gradually replacing the Malays as a trading group.

The traditions of inter-island, coastal, and riverine trade in Borneo before the eighteenth and nineteenth centuries are difficult to reconstruct. There is no known written record of the trading transactions between the traders and the inland tribes. The main configuration of such basic economic activities which have persisted over time can only be broadly sketched.

The Malays, who once played the role of traditional Bornean traders, inhabited the coastal areas of strategic river-mouths and the confluences of rivers and their major tributaries. At a local level, trade was conducted between the river-mouths and the up-river hinterlands of Borneo. These river-mouths were in turn linked with neighbouring ports and islands as part of an archipelago-wide system of trade and exchange.

The Malays evolved a special relationship with the up-river animist natives. The coastal Malays used their connections and knowledge of the outside world to advantage. For example, some Malays claimed flimsy connections with Brunei, the theoretical suzerain,[26] in order to exact tribute in the Sultanate's name. One activity in which the upper strata of Malay society was to excel was trade. And trade was combined with raiding expeditions, the speciality of up-river Ibans of the Skrang and Saribas Rivers, who gained notoriety as 'pirates' in the eyes of James Brooke. Trade was linked to political control, for those who could control the movements of people and goods up and down a river, or along the coast, were in a position to impose taxes, tolls, and tribute payments, and charge high prices for essential items like salt and cloth. Politics was inevitably intertwined with trade. For traders to be successful, political machinations were essential, whether these consisted of encouraging the Ibans to raid or of gaining control over strategic riverine points.

Where the Malay traders were powerful enough, as in the First Division, 'forced trade' or negative reciprocity was commonly practised, whereby overvalued and unsolicited goods sent to a particular up-river community had to be accepted and paid for. One observer noted, 'A trader thinks little of asking a pikul of bees' wax for a common English sarong or cloth sufficient to make a *baju*,'[27] and a nineteenth-century traveller described how 'forced trade' was carried out:

The Pangeran Makota, for instance, would send to a Dayak village an invoice of rice, cloth, gongs, iron or salt at a price from six to eight times their real value, and in payment, he would demand, at one-eighth of its value, any produce the Dayaks possessed. The profits from these transactions sometimes reached as high as one thousand five hundred percent of the amount invested. If the Dayak declared himself wholly without property, starving and unable to pay, the reply would be; 'then give me your wife, or your child', and there was sufficient power behind the demand to enforce payment in some form. If a clan stubbornly refused payment, it would be threatened with an attack from a more hostile clan, and in one way or another, the Malays managed to keep them in abject poverty.[28]

The political skills of the Malay trader in having a retinue of sympathetic Ibans who would raid neighbouring communities at his behest or encouragement contributed much to his commercial success. Ibans raided their neighbours simply because they liked warfare and, with it, the prospect of collecting heads. This trade–raiding relationship was very evident in the Second Division. The Malays and Ibans of Banting in the Lower Batang Lupar fought together against the Malays and Ibans of the Saribas River.[29] Political loyalties lay with geography and the ability of the Malays to command a following.

Elsewhere in Sarawak, where trading and raiding were not quite as much an inclination of the Malays and where other natives were not to be easily coerced, Malay traders were more careful in their trading transactions. In the Baram River area, the Malays did not appear to exercise a political role:

A trader from the coast, whether Malay or Dayak, when he ascends the river with his small boat, stops at an assigned place, and sends word of his arrival, with a description of his tribe, object and cargo, to the chief, who orders a party of his people to bring the goods to the village; and though this may be four or five days' journey in the interior, it is done without the slightest article being pilfered. The merchant entirely loses sight of his wares, which are carried off by the Kayans, and he is himself guided by a body of the superior members of the tribe. On arriving at the village, a house is allocated for his use, his merchandise is placed carefully in the same habitation, every civility is shown him, and he incurs no expenses. After a few days' residence, he moves his goods to the mansion of the chief, the tribe assemble, and all the packages are opened. Presents are made to the headmen, who likewise have the right, according to their precedence, of choosing what they please to purchase; the price is afterwards fixed, and engagements made for payment in beeswax, camphor, or birds' nests. The purchasers then scatter themselves in the woods to seek the articles, and the merchant remains in his house, feeding on the fat of the land for a month or six weeks, when the engagements being fulfilled, he departs a richer man than he came; his acquired property being safely carried to his boats by the same people.[30]

From this account, it can be seen that elements of mutual reciprocity are evident. The trader maintains goodwill with his clients by giving presents to the leaders of the longhouse, and he, in turn, is provided with accommodation and food. This passage exaggerates somewhat the profits to be gained from trade. Nevertheless, it was an appealing occupation for many Malays. The Malay proclivity for trading had been observed by Hugh Low, a future Resident of Perak:

... their taste for the pursuit of trade is quite a passion, and during their early life, they look steadily and anxiously forward to the time when they shall be able to indulge it with profit.... The more adventurous Malay with some capital would build a boat and would find fellow traders to sail with him, sharing the costs of running it and being allowed to have a certain quantity of the tonnage of the vessel for their own purposes.[31]

In addition to trading with the interior animists, the Malays also engaged in archipelago-wide trade with places beyond Borneo.

When Singapore was established as a regional entrepôt, trading vessels sailed regularly from Borneo with jungle produce for that port. In 1824, forty *prahu* from Borneo called at Singapore.[32] There was a string of Malay trading settlements, along the coast of Borneo on the South China Sea side, from Tanjong Datu at the northern end to the towns of Sambas, Mampawa, and Pontianak on the northwestern side. The coastal trade was in the hands of local Malays and Bugis who sailed in *prahu* of 800–1,200 *pikul* burden, carrying from forty to sixty men each. The boats made two voyages a year. Their cargo consisted of sago, pepper, camphor, beeswax, bird's nest, tortoiseshell, and pearls. Each consignment, according to the individual size of the vessel, was worth from 2,000 to 8,000 Spanish dollars annually.[33] The cargoes were then sold off to Chinese merchants in Singapore in exchange for blue and white Madras cloth, iron, steel, Chinese gold thread, Bengal chintzes, European chintzes, and long cloths, all to be retailed in the interior of Borneo.

It is estimated that Singapore's trade with this part of Borneo was in the region of 60,000–70,000 Spanish dollars annually. The mining settlements of Sambas, Mampawa, and Pontianak sent some of their gold-dust exports to Singapore. From Sambas, about fifteen to twenty *prahu* visited Singapore every six months, bringing 50–60 *bunkal* of gold dust. About twenty to thirty *prahu* from Mampawa and Pontianak sailed to Singapore annually, carrying shipments of gold dust, diamonds, tin, and rattan. The varied cargo from these places was worth between 2,000 and 20,000 Spanish dollars per load.[34] Pre-Brooke Sarawak participated in the wider archipelago trade as well. Up to 120 boats of 'large tonnage' were estimated to enter the Sarawak River each year,[35] and some of these boats could have been owned by Malays. As the hinterland of the Sarawak River, Bau, was rich in gold, Malay *nakoda* sent about a *pikul* of non-*kongsi* gold each year to Singapore.[36]

Moving into the period after the establishment of the Brooke Raj, oral accounts can be relied upon to show the resilience of Malay trading activity at an island-wide level, stretching from Pontianak in

Kalimantan to Tanjong Datu in Sarawak, and ending up in Brunei. The dynamics of Malay commerce lasted until the close of the nineteenth century when the Malay traders could no longer compete with the Chinese. The account below relates the experience of one Malay trader and reveals the forces at work which gave the Malays fame as skilled boatmen and traders:

> Gone are the days of the famed Haji Sleh, son of Pangawa Pontianak, a nobleman from the place of that name in southern [Indonesian] Borneo. Pangawa Pontianak had a boat with which he traded from the south round Tanjung Datu to the Sarawak river. One day he stayed too long at the receiving end and married a Sarawak Malay lady. His son, Haji Sleh, grew up under the wings of his wealthy father and became an important person in his own right.
>
> ... Haji Sleh had a great fancy for sailing and trading, like his father, Pangawa Pontianak; and his friends came readily to ask to be taken with him on these voyages. The skill with which he manoeuvred his sailing boat was admired by all. At every port he called, he was held in esteem.
>
> ... His schooner, named *Sri Lambir* was built of wood, three masted and carried a gross weight of 14 *koyans* (16 tons) with a complement of 13 persons, Haji Sleh being the master. For many years he cruised the coastal seas from Tanjong Datu to Brunei. The arrival of his schooner in every port was hailed with joy, everyone looked to the new goods she brought.
>
> ... Haji Sleh sailed in any weather. On one occasion, the last with *Sri Lambir*, while the sea was calm, he departed for Pontianak. When he was off Sematan, close to Tanjong Datu, the schooner sank in 12 fathoms of water. The crew swam and managed to swim ashore. The cause of the wreck was said to be sudden gust of wind striking a jib-boom near the vortex of a water spout. Some said this was an act of the god of the sea, to humiliate a haughty man. At the time of the sinking, Haji Sleh was seen holding his sword, brandishing it left and right at unseen objects, walking to and from the stern, before his schooner finally capsized.
>
> This story had a happy ending. Haji Sleh and his crew presented themselves to his relative, the then Datu Bandar in Kuching. He took their story to the Rajah. A sum of $100 was paid to help them. Haji Sleh went back into business, this time with a smaller boat, a *bandong* capable of carrying one and a half tons.[37]

This narration of Haji Sleh's life[38] was given some time towards the end of the nineteenth century. A number of important elements emerge from the reconstruction of this oral account. The active role of Malay traders in Borneo is emphasized. The mixed origins of the Malays in Sarawak are pointed out. Haji Sleh's father, who was from Pontianak, married a Sarawak Malay lady whom he met when he sailed up the Sarawak River, possibly all the way to Kuching.

Following his father's footsteps, Haji Sleh traded in a 16-ton sailing schooner which went from one end of the coast, Tanjong Datu, to the north, stopping at Brunei. As owner and master of the vessel, Haji Sleh had a crew of thirteen persons who were likely to have been minor partners and workers. Owning a boat and being able to master nautical skills were signs of social status for the Malays. The schooner was involved in coastal trade, stopping at various stations along the way, and did not appear to have been engaged in up-river trade. This account stresses the important role played by trading boats in a large island criss-crossed with rivers, where water—not land—linked places together.

Malays in pre-Brooke Sarawak were involved in an extensive trading network encompassing down-river ports and up-river hinterlands, ports of neighbouring territories, and—beyond Borneo—regional entrepôts like Singapore. After the Brooke state was established in 1841, Malay commerce was to face a gradual demise when Chinese sojourners and *émigrés* began to engage in trade in the main towns and in rural up-river areas, in an expansive web which rivalled that of the Malays. This form of Chinese pioneering experience was a continuation and a reassertion of an earlier Chinese–indigene pattern of economic exchange which could be traced back to the eighth century at least, on the strength of archaeological evidence.

1. Wang Gungwu, 'The Nanhai Trade, A Study of the Early History of Chinese Trade in the South China Sea', *JMBRAS*, Vol. 31, No. 2 (June 1958), p. 21.
2. Lucas Chin, 'Archaeological Work in Sarawak', *SMJ*, Vol. 23, No. 44 (1975), pp. 1–7.
3. Tom Harrisson, 'A Niah Stone-age Burial-jar, c-14 Dated', *SMJ*, Vol. 16, No. 32–3 (1968), p. 65.
4. Tom Harrisson, 'Gold and Hindu Influences in West Borneo', *JMBRAS*, Vol. 22, No. 4 (September 1949), pp. 33–110; and Tom Harrisson and Stanley J. O'Connor, 'The Tantric Shrine Excavated at Santubong', *SMJ*, Vol. 15, No. 30–1 (1967), pp. 201–22.
5. Tom Harrisson and Stanley J. O'Connor, 'The Prehistoric Iron Industry in the Sarawak River Delta: Evidence by Association', *SMJ*, Vol. 16, No. 32–3 (1968), p. 49.
6. Tom Harrisson, 'Ceramics Penetrating Central Borneo', *SMJ*, Vol. 6, No. 6 (1955), pp. 549–60.
7. Tuton Kaboy and Eine Moore, 'Ceramics and Their Uses among the Coastal Melanau', *SMJ*, Vol. 15, No. 30–1 (1967), pp. 10–29.
8. Lucas Chin, *The Cultural Heritage of Sarawak* (Kuching: Sarawak Museum, 1981), p. 13.

9. According to Jan Wisseman Christie, P'o-ni—on the basis of archaeological evidence—was centred on Santubong, although drawing upon feeder ports stretching to Brunei. See Jan Wisseman Christie, 'On Po-ni: The Santubong Sites of Sarawak', *SMJ*, Vol. 34, No. 55 (December 1985), p. 80.

10. W. P. Groeneveldt, *Historical Notes on Indonesia and Malaya, Compiled from Chinese Sources* (Jakarta: C. V. Bhratara, 1960), pp. 109-10.

11. Ibid.

12. Chau Ju Kua, *Chu-fan-chi*, translated by Friedrich Hirth and W. W. Rockhill (St. Petersburg, 1911), p. 156.

13. Tom Harrisson, 'Bisaya: Borneo-Philippine Impact of Islam', *SMJ*, Vol. 7, No. 7 (1956), pp. 43-7.

14. Tom Harrisson, 'Interesting Bronzes with Some Ceramics Parallels from Brunei and Sarawak', *SMJ*, Vol. 12, No. 25-6 (1965), pp. 143-7.

15. Cited by Tien Ju-Kang, 'The Early History of the Chinese in Sarawak', unpublished supplement to *The Chinese of Sarawak: A Study of Social Structure* (London: London School of Economics and Political Science, 1953), p. 12.

16. Cited in ibid., p. 5. Also see J. R. Logan, 'Notices of Chinese Intercourse with Borneo Proper prior to the Establishment of Singapore in 1819', *JIAEA*, Vol. 2 (1848), p. 612.

17. Daniel Beeckman, *A Voyage to and from the Island of Borneo* (London: Dawsons, 1973; first published 1718), p. 91.

18. Ibid., p. 89.

19. Tien, 'The Early History of the Chinese in Sarawak', p. 7; Logan, 'Notices of Chinese Intercourse with Borneo Proper', p. 613.

20. Lenard Blusse, 'Chinese Trade to Batavia during the Days of the V.O.C.', *Archipel*, Vol. 8 (1979), pp. 195-213.

21. Ibid., p. 210.

22. Thomas Forrest, *A Voyage to New Guinea and the Moluccas from Balambangan, including an Account of Magindano, Sooloo and Other Islands* (London: G. Scott, 1779), p. 381.

23. Logan, 'Notices of Chinese Intercourse with Borneo Proper', p. 615.

24. James Warren, *The Sulu Zone, 1768-1898: The Dynamics of External Trade, Slavery and Ethnicity in the Transformation of a Southeast Asian Maritime State* (Singapore: Singapore University Press, 1981), p. 5.

25. Ibid., p. 6.

26. See Edward Parnell, 'The Tributes Paid in Former Days by the then Dependent Provinces of Sarawak', *SMJ*, Vol. 1, No. 1 (February 1911), pp. 125-30.

27. 'Dalton's Papers on Borneo', in J. H. Moor (ed.), *Notices of the Indian Archipelago, and Adjacent Countries*, 2nd edition (London: Frank and Cass, 1968), p. 15.

28. William Hornaday, *Two Years in the Jungle* (London: K. Paul, Trench, 1885), pp. 342-3.

29. Robert Pringle, *Rajahs and Rebels* (London: Macmillan, 1970), p. 62.

30. Rodney Mundy, *Narrative of Events in Borneo and Celebes, down to the Occupation of Labuan: From the Journals of James Brooke Esq.*, Vol. 1 (London: John Murray, 1848), pp. 263-4.

31. Hugh Low, *Sarawak; Its Inhabitants and Productions: Being Notes during a Residence in that Country with H.H. The Rajah Brooke* (London: Bentley, 1848; reprinted Singapore: Oxford University Press, 1988), pp. 135-6.

32. Moor, op. cit., p. 4.

33. Ibid., p. 13.
34. Ibid.
35. Low, *Sarawak; Its Inhabitants and Productions*, p. 134.
36. Ibid., p. 26.
37. Tom Harrisson, *The Malays of Southwest Sarawak before Malaysia: A Socio-ecological Survey* (London: Pall Mall, 1970), pp. 232-3.
38. Tom Harrisson and George Jamuh, 'Niah: The Oldest Inhabitant Remembers', *SMJ*, Vol. 7, No. 8 (December 1956), pp. 454-63.

4
Chinese Traders in the Lupar, Rejang, and Baram River Basins:[1] The Origins and Pioneers of *Ulu* Trade

THE expanding frontiers of Sarawak provided opportunities for the Chinese to establish a more permanent presence by building shops, usually close to the security of a Brooke office or fort. While many Chinese took advantage of the safety afforded by a Brooke representative, there were others who traded beyond the bounds of Brooke authority. Rural riverine trade was exemplified by the proliferation of bazaar shophouses as well as by up-river peddling undertaken by boat hawkers.

The location of bazaars and the conduct of up-river trade were influenced by the flow of rivers and the products of their hinterlands. The Chinese pioneers who took up trading as an occupation had the arduous task of encouraging the various native groups inhabiting the river basins to procure the various products of the jungle for them—products such as the more ordinary gums and rattans, as well as more exotic items such as rhinoceros horns, bezoar stones, and bird's nest. In return, the traders offered what the indigenes wanted—luxury goods such as ceramics and beads, and utility items such as cloth and cooking utensils, ornamental paraphernalia, and essential foodstuffs in the form of salt and canned provisions. It was a reciprocal activity, the expansion of which was as much due to the trading skills of the Chinese as to the needs of the indigenes and their industry in procuring the produce of the jungle.

The Origins and Expansion of Trade

The Batang Lupar Basin

Trade in rural riverine Sarawak generally expanded with the political frontiers of the state. The Batang Lupar basin was ceded to the Rajah by the Sultan of Brunei in 1853. An established European presence provided security and encouragement, though of course there were Chinese whose presence in the district preceded Brooke rule.

A physical description of the rivers will give an indication of pioneering conditions in a riverine frontier area, and will show how

Map 5 The Lupar River Basin

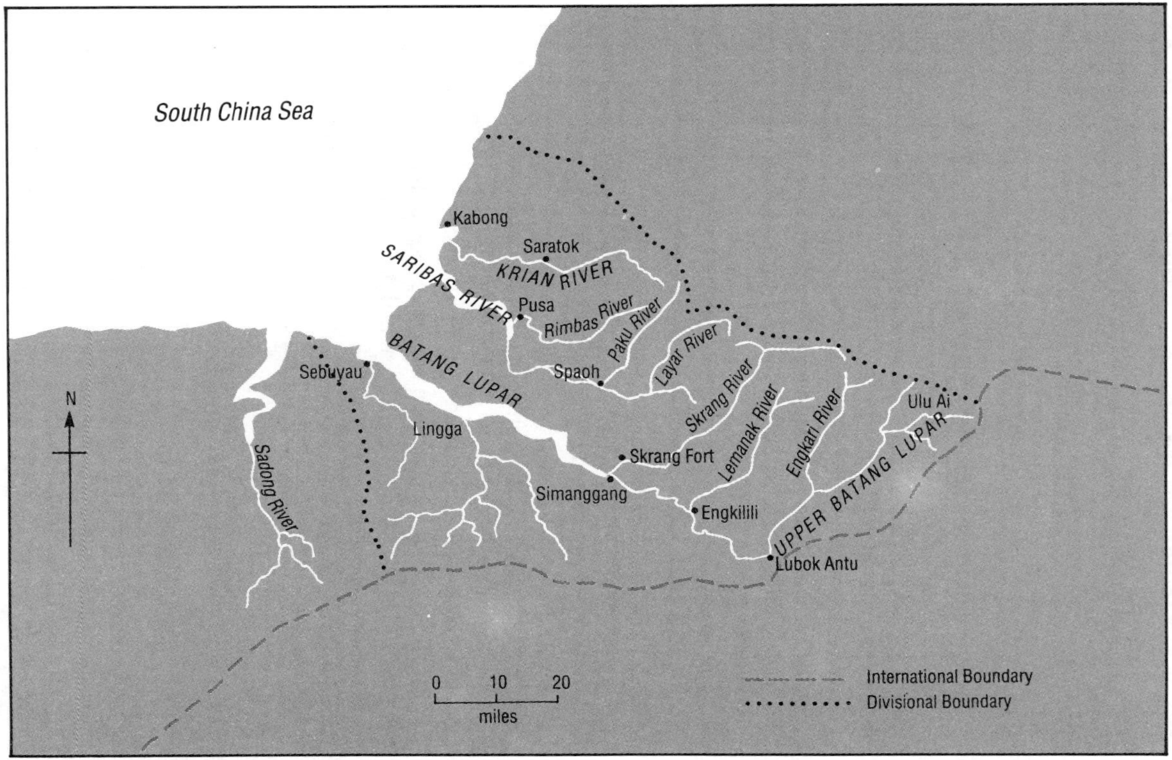

the bazaars, the symbol and mainspring of riverine trade, originated and developed. The rivers in this Division are the Batang Lupar, the Saribas, and the Krian. These three rivers are not long by Sarawak standards, but they are situated in the midst of Iban country (see Map 5). The Batang Lupar, only 120 miles long, rises in the watershed with Kalimantan, and runs south-west to Lubok Antu through hilly country where it is known as the Ulu Ai, then north-west to the sea, through a plain to Simanggang (Sri Aman) as the Batang Ai, and then through low-lying swampy country as the Batang Lupar. The river runs from mountain to hill, from hill to plain, and finally through swamp, the basic topographical features of Sarawak. The Saribas is an 85-mile-long river, rising in Bukit Sadok where it is called the Layar. The Krian River, which is 60 miles long, rises 18 miles east of Saratok, and then runs through low-lying land to the sea at Kabong. The area drained by these three rivers did not constitute an inviting region though it was known to contain *damar*, *gutta*, *jelutong*, and *engkabang*. This lack of resources stood in stark contrast to mineral-rich Bau. Nevertheless, there were Chinese who felt motivated to try and eke out a livelihood as traders in this poor and dangerous region.

Before the advent of Brooke rule, there were no permanent shops, conditions being unsafe for trade. But this is not to imply that there were no trading arrangements of any kind. According to the oral accounts of the Ibans of the Saribas River, two generations before the arrival of James Brooke, Unggang (also known as Lebor Menoa), an Iban 'pioneer raider', is said to have granted permission to Chinese traders from Kuching to trade along the river. These traders were to always fly a white flag on the masts of the boats. They brought cooking pots, earthenware, shell armlets, bracelets, red beads, cowrie shells, and bells of different sizes. The Ibans paid for these goods with padi. One red bead cost them a *passu* of padi. At this time there was still no demand for gums or resins. The Chinese cautiously stayed in and traded from the security of their boats.[2] Local Chinese oral reports maintain that before James Brooke built Fort James, the first of his forts in the Second Division, in 1849, at the mouth of the Skrang River (a tributary of the Batang Lupar), to check the Iban 'pirates', a few Teochius were already farming the land and were engaged in trading in Iban longhouses.[3] When the fort was built, its presence could have given encouragement to the Teochius to build the first five shops there.

The low-lying land around the Skrang, with its propensity to flooding, caused the fort to be shifted in 1864 to Simanggang, 58 miles from the mouth of the Batang Lupar. There, it was renamed Fort Alice.

The traders followed suit and, from this nucleus, the Simanggang bazaar had grown to forty-six shops by 1872.[4] Fort Alice overlooks the present bazaar. It stands on a hill with a commanding view of the river and is well positioned to observe boats either going up- or down-river. In its pioneering days, the Simanggang bazaar faced the river directly, adjacent to the fort. Today, the bazaar has of course expanded inland. Along the river-front there is still a row of corrugated zinc-roofed shophouses with hardwood framework. This was the spot on which stood the original shophouses, burnt down in 1928. The original structure was a row of 'open-fronted' shops[5] (this being a typical shophouse style for the period), made of *attap* roofs and timber walls, two storeys high. The bottom storey was the trading place, and above were the sleeping quarters and store. The individual shophouses all looked alike and stocked the same goods. A Brooke official commented: 'They all seemed to stock the same goods—tinned salmon, sardines, and brown, dirty-looking biscuits, jars of weird preserves, rolls of Turkey red and flowered muslin. They all had the same smell too—a conglomeration of pungent spices, mildew and mustines.'[6]

The bazaar is now surrounded by three Malay kampongs, one immediately to the left, the second to the right, and the third one across the river. Right in front of the shops are jetties, where boats load and unload their freight. Up-river were the Iban longhouses, in which the main clientele of the bazaar were to be found. Down-river from Simanggang was Lingga, 30 miles away. It had at least twenty-three shops in 1883.[7] Further down towards the coast, at the mouth of the Lupar, 58 miles from Simanggang, was Sebuyau. Both Lingga and Sebuyau had demographic features similar to Simanggang, being sited along the river, and situated in the midst of Malay kampongs. As one went further down-river, it became more of a Malay–Muslim area, with Malay kampongs and individual houses, and occasional Iban longhouses. Lingga was indispensable to Simanggang in the nineteenth century, being a port of call for Chinese launches running between Simanggang and Kuching, and an anchorage for ships too large to travel up the Batang Lupar and for vessels awaiting the passage of the tidal bore just above Lingga.

Simanggang, with its central location, was the hub of trading activities on the Batang Lupar. Credit, goods, and transportation extended outwards from the divisional capital. Simanggang was a Teochiu bazaar. It was no coincidence that the bazaars down-river, Lingga and Sebuyau, were also Teochiu in character. Up-river shophouses in Engkilili and Lubok Antu were Teochiu-owned, too.

The Teochiu pioneering spirit led to the building of these smaller bazaars. Kinship and credit ties were a second reason which linked the outer-Simanggang shops to Simanggang.

There was an increasing demand in the 1870s for jungle gums and resins, sought after for industrial purposes. To get to the very source of these jungle products, traders sailed up the river from Simanggang to Engkilili and Lubok Antu. Lubok Antu is 30 miles up-river from Simanggang. The Batang Lupar meanders slowly to Lubok Antu, and there are no hazardous obstacles except at the mouth of the Lemanak (below Engkilili) where there are rocks and a shallow gravel bed. Launches need the aid of tide or flood-water to pass these obstacles. Half-way between Engkilili and Lubok Antu, there are several small rapids passable with flood-water. In 1872, Teochius travelled up to Lubok Antu to build shops[8] when it was learned that *gutta*, beeswax, and rattan were available. Engkilili, 15 miles up-river from Simanggang, did not have a bazaar constructed until 1920; the bazaar still stands today, a row of timber-framed, two-storey, zinc-roofed shophouses directly facing the river. Before the shops were built, this place was visited by trading barges (locally called *jala*). According to oral tradition, the river boats, 50 to 60 feet long, had to be hauled upstream by crews of eight persons[9] as there were no winds, and there were rapids to negotiate below Engkilili. The local name for Engkilili was *Tueh Tau*—'top of the steps' in the Teochiu dialect—because the traders had to use wooden steps to get ashore.

The mouth of the Batang Lupar is only 3,520 yards (2 miles) wide; at Simanggang, the river is 320 yards wide; and at Engkilili and Lubok Antu, it has a width of 50 yards. In the nineteenth century, it was not considered a rich river basin when compared with the Rejang and Baram. Being an 'original' Iban heartland, the area had lost most of its fertility through continuous burning, farming, and the extraction of forest produce. The hinterland being famished, the bazaars did not prosper. The other two river basins of the Division are the Batang Saribas and the Batang Krian.

Betong was the only major bazaar along the Batang Layar, a branch of the Saribas, an autonomous river basin which had no kinship ties and economic links with the Batang Lupar, even though only 12 miles of land separated Betong from Simanggang. 'Layar' was the Iban name for the Saribas River, while 'Saribas', being the Malay name for the lower main river, was applied to the entire river system by the Brookes.[10]

Betong is 50 miles from the mouth of the Saribas. It had twenty *attap*-thatched shops in 1871.[11] As with Fort Alice at Simanggang, the shops at Betong were overshadowed by Fort Lili, a Brooke station built in 1858. The bazaar was situated inland from the river banks because of the danger of the river flooding its banks. At present, there is a Malay kampong along one of the river banks. Iban longhouses are located further up the headwaters of the Batang Layar, along the Paku tributary, and further down the main river. Betong shared features similar to those of Simanggang, with Malay kampongs close at hand and Iban longhouses further up- or down-river. For almost half a century, Betong was the only major bazaar along the Saribas. There was little incentive for the expansion of trade in this isolated river basin surrounded by rugged hills. Chinese junks from Kuching took five to six days to reach Betong and, with low water and little wind, could take as long as ten days to reach their destination in Betong.[12]

It was therefore hardly surprising that the next bazaar of any importance was not built until 1916. Spaoh is situated at the mouth of the Paku tributary, 5 miles down-river from Betong. The shops in Spaoh bazaar were built to take advantage of the rubber boom in the Paku tributary.[13] The boat hawkers operating from Betong had requested permission from the Resident to set up some shops. The Resident's reply was, ' ... as their trading boats are kept stationary on the river bank and they purchase all their goods from the Betong bazaar, I saw no harm in this, and four shops will shortly be erected here.'[14]

Proceeding further north, one finds Saratok, situated in the Kalaka district drained by the Batang Krian and 24 miles from the sea. It is not known when the bazaar was built, nor how many shops it originally possessed. The shops must have been in existence for some time, as it was stated in 1885 that 'a new bazaar was to be erected, the present one being in a dilapidated condition'.[15] Saratok grew to be a bazaar rivalling Simanggang, and in 1941 had seventy shops.[16] Its relative prosperity could probably be attributed to its peaceful location. Traders in the Batang Krian enjoyed considerable autonomy, building shops and trading wherever they liked, until in 1890, D. J. S. Bailey, Resident in Simanggang, ordered them to build shops in Roban, along the Seblak tributary of the Krian, 15 miles from the river-mouth.[17]

The bazaars along the Lupar, Saribas, and Krian Rivers had been constructed between the 1860s and the 1870s to take advantage of trading opportunities, albeit limited in scope. The individual personality of each of the three rivers defined the character of the bazaars—and of trade, the *raison d'être* for the presence of the Chinese pioneering traders in this riverine frontier.

The Rejang River Basin

As in the case of the Second Division bazaars, the addition of the Rejang River basin to the dominion of Charles Brooke in 1861 gave encouragement to Chinese *émigrés* to trade along that river. The Rejang River, 350 miles long, is the longest in the state. It is navigable by ocean-going vessels as far as Kapit, 160 miles from the river delta. The river-mouth is tidal and muddy, traversed by many navigable channels. The river is widest at its mouth where it measures 5,280 yards (3 miles) across, then it tapers to a width of 1,320 yards (three-quarters of a mile) at Sibu, 250 yards at Kanowit, and widens again to between 300 yards and 400 yards at Belaga.

The image of a lush, tropical frontier best fits the Rejang River, with its imposing length and width and a rich jungle hinterland. In this respect, the Rejang River differs markedly from the shorter rivers of the Second Division. However, the manner in which the pioneering Hokkiens and Teochius traded and established bazaars in the Rejang River basin was similar to the way in which the Second Division pioneers traded along the Lupar, Saribas, and Krian Rivers, with shophouses built adjacent to Brooke stations, either to take advantage of Brooke security or at the behest of Brooke officers. Just as Simanggang acted as the commercial riverine centre for the Batang Lupar district, so too did Kanowit assume that role for the Rejang River basin.

According to oral accounts, junks sailed up and down the Rejang River, retailing salt, crockery, and cloth to the Ibans in exchange for jungle products, before the first influx of permanent Hokkien settlers from Singapore arrived in the 1870s at Kanowit, 60 miles from the coast.[18] Before the Hokkiens arrived and built their shops, a bazaar occupied by Malay traders from Kalimantan and the Rejang coast was in existence. In 1865, a traveller described these early Kanowit shophouses thus: '... the bazaar was principally occupied by Malays intermixed with a few Chinese; the houses are built upon piles fifteen feet high, with a broad verandah in front, and a narrower one behind'.[19] A Chinese resident at Kapit, further inland up the Rejang River, described how, in the 1930s, the shophouses of that bazaar had a 10- to 15-foot-wide veranda in front of the shops for storing rubber and other jungle products.[20] From a distance, the shophouses

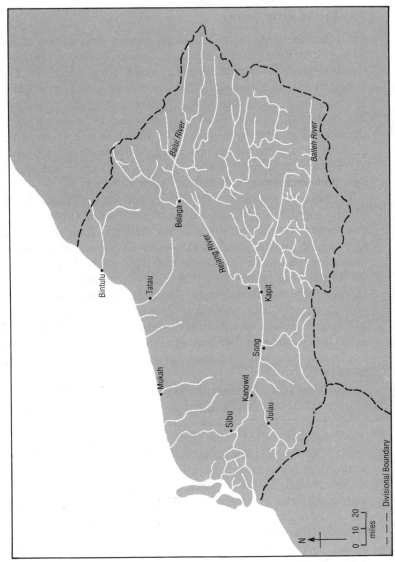

Map 6 The Rejang River Basin

would have resembled an Iban longhouse on stilts. The reason the shophouses were raised could have been to lessen the likelihood of flooding, for the river level could rise very suddenly after a torrential downpour.

The arrival of Hokkien merchants in Kanowit put an end to the trading activities of the Malays and, by the 1880s, the Hokkiens were fast becoming a stable and settled community. Some of Kanowit's present-day Chinese can trace their roots back to this pioneering group of the 1880s.[21] There is no indication at all as to the relative size of the Kanowit bazaar community, only that oral recollections and written records attest to the fact that it was bigger than down-river Sibu, which is in the 1980s the biggest town and the administrative centre of the Division.

Sibu in 1871 had sixty shops[22] and was growing in size and importance as a centre of trade and administration. By 1880,[23] it had surpassed Kanowit in size and importance, and it seemed likely that, by then, part of Kanowit's trade was being diverted through down-river Sibu. In 1890, the Resident of the Division was based in Sibu; this was indicative of the growing importance of Sibu relative to Kanowit. In 1875, up the Rejang River, a fort was constructed at Nanga Balleh, at the confluence of the Balleh tributary and the Rejang River, 55 miles from Kanowit. A bazaar was soon built there, in conformity with the fort–bazaar pattern:

A bazaar will spring up around the fort (several houses are already in the course of erection) and it will speedily become an entrepot for the different tribes of the interior to carry on their traffic, and where they will bring their jungle produce to barter for necessaries, which John Chinaman knows so well to supply.[24]

Two years later, this location had twenty shops.[25] Within a year, however, there was less optimism about the Balleh bazaar:

There is not much appearance of advancement in the Balleh bazaar, but should an improvement take place, the station will have to be shifted five miles further down the river, to Kapit, where there is safe anchorage; the Chinese too, are very desirous of this removal; this was the place originally decided upon.[26]

There was an obvious reason for this proposed shift. Iban migration to the Balleh and their zeal for raiding and head-hunting had made conditions unsafe for their weaker neighbours—the Ukits, Penans, and others—and for trade. The government post at Ngemah and the presence of a government steamer did not stop this wave of Iban migration which had reached as far up-river as the mouth of the Mujong River, nor did it stop head-hunting. Defiant Ibans had retreated to Bukit Batu, a limestone crag at the headwaters of the Mujong tributary of the Balleh River, where they fought two government expeditions before surrendering in 1881. After this, the whole of the Balleh River was cleared of Ibans and declared closed to further settlement.[27] By this time, the government station had been shifted 5 miles down-river to its present-day site at Kapit (the move occurred in 1875). The traders followed suit and, by 1888, Kapit boasted eighteen shops with seven others under construction.[28]

The government station furthest up the Rejang River was at Belaga, some 95 miles away from Kapit. The post was established in 1884, and it soon provided the necessary impetus for the building of a Chinese and Malay bazaar. Before that, itinerant traders were described as having 'traded for years among the Kayans in the desultory manner by going from place to place'.[29] The fort provided security, and a proposal to dismantle it several years later was a cause of considerable disappointment to the traders.[30] Charles Brooke eventually decided that the government outpost would remain.[31] In 1893, Belaga had fourteen shops, with reports of fifteen more to be built, ten to be owned by Chinese traders, and five by the Malays.[32] When a fort was built in 1892 at Song, between Kanowit and Kapit, it too attracted Chinese shopkeepers. The following year, it was reported that 'seven new shops were in the course of construction'.[33]

From Kanowit, the traders spread down-river to Sibu and up-river to Kapit, Song, and Belaga. Shophouses were built far up-river to be nearer the rich sources of the jungle products. A rich pioneering frontier similar to the Rejang River basin was the Baram River.

The Baram River Basin

This riverine basin was ceded to Charles Brooke in 1882 and, as with the other river basins, bazaars sprang up in conformity with the fort–bazaar pattern. As in the case of the Rejang River basin, the Baram riverine frontier was similarly well endowed with jungle products. The Baram River, 250 miles long, rises in the Kalimantan watershed, cutting through hilly country and meandering considerably in its course towards the low-lying country around Kuala Baram, on the coast, 14 miles north of Miri. Its width at the mouth is 1,760 yards (1 mile), narrowing progressively up-river to widths of 200–400 yards between the mouth of the Bakong and Marudi, 150–200 yards at Long Lama, and 100 yards at Long Akah. Not

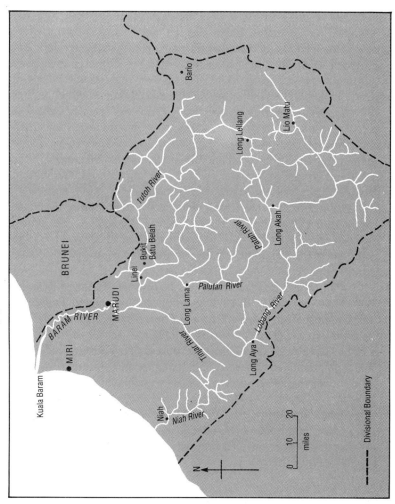

Map 7 The Baram River Basin

unlike the Batang Lupar with its tidal bore, communication in the Baram is hampered by a sand bar which runs from 1 to 3 miles out from the river-mouth with a depth of less than three feet of low water. Only ships which can cross the bar can proceed to Marudi. Boats can reach Lio Matu, where the first rapids are below the mouth of the Palutan, 15 miles up-river from the last Chinese bazaar at Long Akah. A decade before Rajah Charles took over the Baram River concession in 1882, the *Sarawak Gazette* had viewed the development prospects of the Baram district negatively: ' . . . the importance of the Baram in its present state may be over-rated in the minds of many; its exports at the most, do not exceed $2,000 monthly, and the interior inhabitants scarce have learnt yet how to work the produce of the jungle; in fact, trade, as it now stands, is uncertain and irregular.'[34] This disparaging comment could have been due to the fact that the Sultan of Brunei still retained nominal control of the river, and Brunei Malays were actively trading with the Kayans and Kenyahs. Most of the trade was carried out in the Tutoh River tributary where there were many longhouses lining the banks of the river, from Long Linei to Long Melinau. At Kelijiu, the Brunei Sultan's representatives had built an *attap* hut which was the *kubu* and it served as the administrative and trading centre of the Baram district.

In 1882, it was decided that a fort be built at Marudi,[35] 42 miles away from the coast.[36] Like Fort Alice in Simanggang, Fort Hose was built on a hill overlooking the river. The bazaar shophouses were constructed at the foot of the hill. There was a Malay kampong near by. The Kayan longhouses were dotted along the banks up-river. The topographic and demographic characteristics of the fort–bazaar were similar to other stations elsewhere in the Batang Lupar and the Rejang.

A pioneering group of Hokkien traders from Kuching arrived in April 1883, on the *Lorna Doone*. The results of their initial efforts in up-river trade were encouraging. Claude Champion de Crespigny, the Resident, said:

Aban Nipa who was formerly so adverse to Chinese of Kuching coming to trade at this place, is now anxious they should do so, and he and the Kenniahs [Kenyahs] were surprised at the prices named by Ken Wat's men—who sold them superior *blachu* [black cotton cloth] for $2 the *kayu*, they have always been accustomed to pay $3 for *blachu* of an inferior description. . . . A number of Brunei Malays who came down with Aban Nipa were angry . . . saying they would be ruined if our Chinese traded at those

THE ORIGINS AND PIONEERS OF *ULU* TRADE 75

prices and were particularly angry when Ken Wat's men paid without beating down the sum demanded for a pikul of good *gutta*, namely $32.[37]

The Resident continued:

The Kayans and Kenniahs are most anxious to commence trade in earnest with our people, and this is why I particularly desired them to come down—that they might know the price of things.[38]

When Charles Brooke paid a two-week visit to Marudi in 1885, he reported seeing forty-five *attap* shops in the makeshift bazaar and counted about 400 Chinese and Malay traders in Marudi and up-river.[39] The following year, 1886, the Marudi Chinese were organizing to build a permanent bazaar complex. In August of the same year, a petition was sent to the Rajah asking for free transportation of building material on the *Adeh*[40]—the traders explained their desire to build permanent shops. The Chinese also wanted to construct a temple for the *Tua Pek Kong* deity, out of *belian*, on one of the small hills in the jungle about 300 or 400 yards behind the fort. The erection of the bazaar did not begin in earnest until April 1889, when a temple resting on *belian* piles was built at the down-river end of the bazaar.

By then, the building of the new bazaar had become a necessity. An accidental fire in March 1889 had destroyed the existing shophouses. Construction began almost at once. The shops were set on *belian tian* (piles) and covered with *belian* shingles. Each shop had a frontage of 7½ yards.[41] By April 1890, ten shops had been completed. At the end of the year, twenty-nine shops were ready, with a road 30 feet wide in front of the shops and two landing stages for the steamer, the *Adeh*.[42] In 1891, the Chinese in Marudi asked the Rajah to sell to them, for a reasonable fee, the land on which the shops and godowns were built as well as a small plot at the back for making gardens. The bazaar was burnt down again in 1918, and rebuilt.[43]

Using Marudi as a base, the traders were to disperse to other places in search of new opportunities. In 1884, a Marudi merchant, Lai Ngee, sent an agent to Niah to open a shop where 'he disposed of all goods in a very short time ... there was a quantity of *gutta* and other produce at Niah but the Chinese have no money to buy with'.[44] In August 1887, Ah Tau of Chop Guan Ann, a leading Marudi merchant, opened a shop at Batu Belah on the Tutoh River, 38 miles from Marudi.[45] The Tutoh River was the stronghold of Brunei Malay traders.

A government census in 1888 estimated the number of Chinese

men, women, and children in the Baram district at 170 persons.[46] By 1891, at Niah, it was stated that there were twenty-seven Chinese traders eager to build permanent shophouses. The traders, desirous of government security, were waiting for an administrative office to be set up before they built their shops. When a site for the new government station was finally selected on a small hill, the bazaar was laid out on the opposite bank of the river.[47] By 1905, a bazaar block of five shops had been constructed at Long Lama,[48] 48 miles from Marudi, on the winding Baram River, with tall, tree-lined cliffs fronting the river banks and an occasional Kayan longhouse breaking the monotonous scenery. According to the local folklore of Long Lama, the bazaar originated with a partnership between three traders, Chan Boh Siong, Woon Chin Mui, and Sim Ah Lok, who until then had been operating independently and separately at the mouth of the Patah River (38 miles from Long Lama), at Long Akah (50 miles from Long Lama), and at Long Lama respectively.[49] The rapid spread of Chinese traders up-river in the Baram basin was due to the abundant trading opportunities available.

In 1919, the Baram *towkay* were informed that, as from 1920, no Chinese or Malay traders would be able to build a shophouse or trade on a permanent basis at a Kayan *pangkalan* (landing place). This was to discourage the traders from associating too closely with the local people. If the Baram shopkeepers wanted to trade up-river, they had to establish proper bazaars from which they could depart for the various longhouses with their trading goods, but they would only be allowed to stay a maximum of two nights at each place. The Resident, C. D. Adams, decided that they could establish bazaars at Long Lama (which had been built earlier anyway), at Long Akah on the Baram River, at Long Lobang and Long Ayah on the Tinjar River, and at Long Apoh on the Apoh River.[50] Chinese traders had gone right up to the Kayan and Kenyah heartland, a land with a reputation for dangerous river rapids.

Where the origins and growth of bazaars in all three riverine basins are concerned, the overall picture to emerge is one of a proliferation of river-front bazaars, usually situated inland and geared towards the procurement trade flowing down from up-river. Some bazaars in Simanggang, Kanowit and Marudi, because of their ideal locations, became regional centres from which the trading diaspora would spread outwards, down-river or up-river. Brooke policies and the security created by the presence of Brooke forts had a conscious effect of confining the Chinese to the bazaars although the traders—whether in or out of the bazaar—enjoyed considerable autonomy.

THE ORIGINS AND PIONEERS OF *ULU* TRADE

In all three river basins, the individual personality of the rivers and the products of their hinterland determined the expansion of trade. The Rejang and Baram River basins, being sparsely inhabited frontier regions with a plentitude of jungle products, were encouraging to traders. The area surrounding the Batang Lupar and its adjoining rivers of the Saribas and the Krian, being the place of first settlement of the Ibans in the state, had been stripped of most of its resources through swidden cultivation and produce collection. The Batang Lupar, Saribas, and Krian bazaars did not enjoy the growth in trade experienced by the Rejang and Baram bazaars.

Pioneering Life

Eagerness on the part of the traders to get to the source of the products of the jungle was to involve several of them in a unique pioneering experience. This was especially true of the boat hawkers who rowed their boats from longhouse to longhouse.

The experiences of the pioneers were similar in many respects: they invariably had to cope with physical hazards, ranging from accidents to murder and robbery; long boat journeys were inevitable; and contact with the natives to effect economic exchanges was essential; most traders relaxed by indulging in gambling, drinking, and opium smoking.

The salient points in the lives of the traders in each of the riverine basins will be highlighted. Oral accounts will be woven into the text in order to tell the story of the pioneers' past experiences 'from the inside', as it were—for example, those of the ordinary boat hawkers who played a major intermediary role between bazaar and longhouse and who were largely responsible for bringing the natives into contact with external economic forces. If some of the accounts of these boat traders are not written and as their number diminishes over the years, their historical role will probably go unrecorded and pass into oblivion. The oral reminiscences of ordinary Chinese boat hawkers are useful in balancing the written accounts of the Chinese pioneers in the *Sarawak Gazette*: the latter were invariably on the successful and established community leaders. Articles published in the *Sarawak Gazette* were also written from the standpoint of administrative considerations such as the maintenance of peace and order, and, towards this end, there was fairly extensive coverage of accidents, robberies, and murders. This makes the oral recollections, particularly about the more mundane events of life, all the more valuable as a source of social history, revealing the lives of the traders from within.

The pioneering traders in all three river basins were mostly Hokkiens and Teochius, with the exception of a few Chao-anns, Hakkas, and Hainanese, and—in the twentieth century—Foochows as well. Most of the Chinese would have regarded their sojourns as temporary, and their objective would have been to accumulate sufficient wealth to retire in some comfort to their villages in China. Due to the lack of statistical evidence, it is not possible to chart the movement of the Chinese to and from Sarawak; this is a pity as this could provide some indication of the numbers who stayed behind. A few of the pioneers would have returned prosperous to their native villages, others would have stayed on in Sarawak. For the majority—perhaps the vast majority—of the *sinkheh* who chose trading as an occupation, their dreams of amassing riches in the *Nanyang* remained unfulfilled. They would have spent most of their days in Sarawak struggling against adversity and, having no choice, were forced to remain behind in Sarawak in remote riverine bazaars, or up-river in their *attap*-thatched boats.

Those in the bazaars were more fortunate. When they were not busy selling or loading and unloading goods, they faced long days with little to do except wait for boats to come up to the moorings in front of the shops. The most relaxed periods for the shopkeepers occurred during the planting season, when the natives were occupied with their crops, and during the monsoon period at the end of the year when river communication was adversely affected.

The boat hawker had a more difficult time. Most boats were *attap*-roofed, and beneath the roof were sleeping quarters and storage space for provisions and goods. The front and back of the boat would have two long oars each. A typical boat manned by two persons could be anywhere from 10 to 15 feet long, 3 feet wide, and 5 feet high. Boat traders normally operated in pairs, in a standing position, one man rowing in front, and the other at the back. There were, however, others who worked alone in smaller boats. The traders lived, slept, and ate on the boat. At night, the boat, rocked by the movement of the river, was the boatman's bed—unless he spent the night at long-houses, which many did. The boat hawkers bathed in river water, undaunted by the fact that the river also served as a sewer of sorts. The boat traders would get up at the crack of dawn and row until nightfall, for up to twelve hours at a stretch, under a hot sun, their journeys being interrupted by stops at longhouses along the way. If they rowed with the tide and current, their task was made much easier; if not, it could be a body-aching job, their arms having to move the oars continuously. Boat pedlars could be away from the bazaar for periods of a week to a month. They were very much at the mercy of the rivers and their idiosyncrasies, and when contact was established with head-hunting Ibans, these pioneers had to rely on their wits, bravery, and tenacity to survive, taking particular care over etiquette and the establishing of goodwill.

The Batang Lupar Basin

In the mid-nineteenth century, the earliest known bazaar in the Batang Lupar district was the row of five shophouses on the Skrang River, in the vicinity of Fort James.[51]

One of the more successful pioneers in the Skrang area was Tay Sze Heng. Tay moved into Simanggang shortly after 1864, and his family subsequently rose to economic prominence and acquired a high degree of social status in Simanggang. Some idea of the influence of the Tay family may be found in a eulogy in the *Sarawak Gazette* of 1 December 1899 after the death of Tay Soon Heng, grandson of Tay Sze Heng. The grandson was described as having belonged to 'the most prosperous firm of Seng Joo',[52] and was praised for his enterprise in promoting trade throughout the Simanggang district, though only twenty-eight years of age at the time of his death. That Tay Soon Heng was a person of some standing in the community was evidenced by the fact that Resident D. J. S. Bailey, missionary worker Reverend W. Howell, and some Malay headmen attended the funeral.[53] One of Tay Sze Heng's sons, Tay Seng Kim, and another grandson, Tay Lim Thian, subsequently became the *kapitan* of Simanggang during the first few decades of the twentieth century. Tay Seng Kim established excellent rapport with the local Brooke officials in Fort Alice, and was their food and liquor supplier.[54] In 1889, 1894, 1901, and 1903,[55] the family firm of Seng Joo was also awarded the highly lucrative tender for supplying opium.

The interlocking positions of *kapitan*–monopoly 'farmer' (seller) placed members of the Tay family in influential positions, giving them access to Brooke officials and enabling them to wield considerable social and economic power within the Chinese community in Simanggang. It has been said that, in 1910, more than half of the shops in Simanggang were owned or run by members of the Tay clan. The Tay family used its wealth, influence, and power to sponsor the chain-migration of the Tay clan from their native village in China, to Simanggang where the Tays, to this day, remain one of the major surname groups. Local informants have said that the past and present number of Tays in Simanggang is due to the presence of the firm,

Seng Joo. Besides sponsoring fellow clansmen, Seng Joo assisted *sinkheh* by giving them employment, or extending them credit to start trading elsewhere, in Engkilili or Lubok Antu for instance. An elderly resident of Lubok Antu recalled how the Seng Joo firm asked her father to open a shop in the Lubok Antu bazaar around 1900.[56] Some idea of the extent of the wealth of the Tay family is indicated in a notification in the *Sarawak Gazette* of 5 June 1907, which estimated the value of the estate of Tay Ah Eh, who died in China in 1907, at $15,548.[57]

The success of this particular Tay family was atypical of the majority of the *sinkheh* in the Batang Lupar basin. Other than farming, a common form of livelihood for *sinkheh* in most of rural Sarawak was river trading.

Boat hawkers faced many physical dangers in the course of their travels along the Second Division rivers—from rocks, rapids, and sudden high water-levels which could capsize their boats. In October 1897, a boat ran aground when it hit a sandy barrier along the Batang Lupar.[58] An unexpected and rapid rise in the river level after a rainstorm could make conditions hazardous for the boatmen, some of whom could not swim.[59] The following describes a typical boating mishap:

An inquiry was held concerning the drowning of Li Ah How of Roban on 10 February 1895. The deceased's companion, Jap Ah Hin, was brought before the Kabong court on the 17 February, and was charged with failing to report this death. It was elicited that the deceased and Ah Hin got into the boat, the former sitting in the bows and the latter in the stern, and Ah Hin propelled it down the bank into the river. As the banks were high, the bows got under water and the boat upset. Ah Hin tried to save his property and made no effort to assist his companion though he could swim, or could have called out to the people in the houses close at hand. For failing to make a report, Ah Hin was sentenced under the Coroner's order, to the full penalty, in view of his making no attempts to rescue his unfortunate compatriot.[60]

Boats plying the coast were exposed to even greater danger. Bailey, the Resident, complained of the lack of seaworthiness of the Chinese coasters:

The Chinese *bandongs* (junks), though they carry two masts and are allowed to take an unlimited number of passengers, are not under any kind of inspection, and it is an extraordinary thing how long the Chinese owners run these boats in an absolutely rotten state. I have known of at least three which have sunk, and many lives lost, in late years, and all were, according to subsequent inquiry, in an absolutely unseaworthy state. I think all decked boats that carry passengers should be examined once a year and should pay a small fee for their permit to take passengers.⁶¹

In December 1909, a coasting boat sank at sea due to adverse weather conditions and poor judgement on the part of the captain:

When the tides were high and the sea was rough, a new small *bandong* of 7 tons burthen, owned by a Kabong Chinese by the name of Chua Pun, was wrecked on the bar on her second trip and the young owner was drowned. She was on her way to Kuching with a full cargo of raw sago from Daro and some *jelutong* from this place. She tried to cross the bar when the wind was slack with a very high sea running, and consequently she drifted on to the shoal near Beting Marau and stranded. The *sampan* on board had been washed away early in the day and the crew were forced to swim ashore. The sailors managed successfully to get to land, but the owner was lost.⁶²

An occupational hazard which boat hawkers had to put up with regularly was the risk to their personal safety and the likelihood of being robbed or murdered. Nevertheless, the possibility of such dangers did not deter these traders from venturing deep into Iban country. In June 1871, a trader by the name of Ah Tong, who with four other companions lived and traded among the Bugau Ibans of Lingga, reported the death of one of his friends. Tan Le had been killed by Mallang Ibans when he was alone in a boat, trading salt with them.⁶³ Almost forty-three years later, in February 1914, again in Lingga, the local police reported that Tee Kow and Ah Hak, two Chinese boat hawkers in the Lingga River, had been seriously wounded and their goods pillaged. A witness, Timpoh, gave evidence, and it was established that the attack and robbery took place at night whilst Tee Kow and Ah Hak were asleep in their boat.⁶⁴ It appeared that the attack was motivated by robbery.

On the night of 18 April 1929, a boat hawker was shot dead while trading up-river on the Saribas. When his body was brought down for examination, it was found that he had been shot at close range with a shotgun and buckshot. Three officials, the Acting District Officer G. R. H. Arundell, Penghulu Jabieng, and Native Officer Abang Omar, immediately proceeded up-river to investigate, but failed to obtain any immediate clues regarding the attack. All Ibans in the vicinity of the murder were questioned, and suspicion fell on two Ibans. A thorough examination was made of the house where the two lived, and a steel box containing $10 in copper coins was found. Ah Seng, the victim, had had $60 in cash, and this was missing. He had previously been attacked and severely wounded twelve years before in 1917.⁶⁵ This case is similar to the preceding one, with robbery being the motive behind the attack.

The official accounts of pioneering life have emphasized the dangers faced by the traders. Some of these dangers, such as floods and rapids, the traders could do little about. But theft and physical attacks on the traders were a different matter and had arisen out of contact situations between persons of different cultural values. The Ibans' love of heads was one reason behind the attacks. A second reason was the Iban belief that the trader, with his display of goods in the boat, must be a wealthy person; the Ibans did not of course realize that the pedlar obtained these on credit from his bazaar *towkay*. A third possibility was that indebted natives, looking for a way out of their predicament, might decide to get rid of their creditors. Finally, some hawkers invited these attacks when they cheated or took advantage of the natives' less worldly ways. Some oral history recollections of the experiences of the latter-day pioneers will now be reconstructed, which is revealing of the dangers the Chinese confronted.

One remarkable boat trader was a Teochiu, Tay Bak Choon, aged ninety-three when interviewed in Simanggang in 1981. He arrived in Simanggang in 1910 at the age of twenty-two and remained a boat hawker until 1926. Tay had a boat 18 feet long, which he rowed up to longhouses along the Skrang River, and to Engkilili and Lubok Antu up-river on the Batang Lupar itself. The Skrang, a tributary of the Batang Lupar, joins the Batang Lupar 5 miles above Simanggang. It is a short stream, navigable for up to 35 miles—but the short length of the river belies its importance. The Skrang Ibans were the much-feared raiders of the pre-Brooke and Brooke eras, prominent among the raiders being Rentap of the Upper Skrang, one of the most decorated Iban 'rebel' leaders in Sarawak history. On the Batang Lupar, Engkilili is only 15 miles above Simanggang with no navigational obstacles, and Lubok Antu is 30 miles away. Between Engkilili and Lubok Antu there are several rapids. A journey up to the end of the Skrang River might take two days, to Engkilili one day, and Lubok Antu two days. Lubok Antu had a dangerous reputation, as feared as that of the Skrang River, and another of Sarawak's up-river 'rebels', Bantin, is still remembered in the oral traditions of elderly Chinese traders.⁶⁶ For a long time, the Lubok Antu traders dealt cautiously with the Ibans, establishing two ways of trading with them. At night, the Ibans would stand beneath the shophouses which were raised on stilts, very much like longhouses, and they would attract attention and communicate by thumping the floor. During the day, exchange transactions were carried out through an open window, the front door being permanently closed.

The boat trader Tay was operating in waters notorious for raiders and 'rebels'. His companion for his travels was his wife, whom he had married in China and brought over in 1912. It was rare for a Chinese woman to accompany her husband in his trading activities along the river. Tay conducted barter exchange in Iban longhouses, collecting rice, padi and *gutta*, in exchange for cloth, tobacco, sugar, salt, biscuits, and canned food. At Engkilili, where there were no shops until 1920, Tay traded with Hakka pepper planters. Goods were advanced on the promise of payment of future pepper harvests. At Lubok Antu, Tay obtained *engkabang* from the Ibans.

Being a pedlar in a river which could provide no guarantee for his safety, Tay had to rely on his wits and bravery to survive. The boat hawker claimed that the Ibans were afraid of him because they thought he possessed supernatural powers. Recalling a diving contest he had had with some Ibans, Tay said, 'I held my breath while under the water. Water bubbles rose to the surface. The surprised Iban bystanders shouted, "Antu! Antu!" (Ghost! Ghost!)'. To highly superstitious and omen-conscious Ibans, extraordinary performances such as these were well regarded. Tay had also on occasion asked some Ibans to stand on his stomach while he lay on his back on the ground. These feats were performed to convey an aura of supernatural strength to make the Ibans afraid of him. Whenever he visited longhouses, because he had gained a reputation for being a charismatic figure, enthusiastic Ibans would point out to him human heads hanging in the veranda and claim they were '*kepala China*' (Chinese heads), in a bid to dampen his spirits. But Tay asserted that he was never afraid nor did he ever have to prove himself in any confrontation with the Ibans. He said he went unarmed among Ibans who usually had their *parang* (long knives) by their sides. But he treated the natives with respect. He maintained goodwill with his clients by allowing debts to go unclaimed and he offered them small gifts now and then.[67] Tay usually stayed overnight in Iban longhouses in the course of his travels.[68]

A boat hawker who was not as fortunate as Tay Bak Choon in striking awe and fear in the minds of his clients was Tay Ing Boon.[69] On one occasion, he was attacked by an Iban to whom he had just sold tobacco, and when he was interviewed, his face still bore marks of knife wounds. Tay first landed in Simanggang in 1921 and, after a long period of apprenticeship working on a junk and then a motor launch between Kuching and Simanggang, he turned to itinerant trading, going as far as the Undup tributary branch of the Batang

Lupar. (The mouth of the Undup is 3 miles above Simanggang, running for about 18 miles to the watershed with Kalimantan.)

When interviewed, Tay Ing Boon was a slight, mild-mannered person with shaggy whiskers. He certainly lacked the presence of the other Tay who, though ninety-three years old when interviewed, had the appearance of a seventy-year-old man. When interviewed, Tay Bak Choon still looked strong, with excellent physique; he was barechested and wore white shopkeeper's trousers with a wide belt around the waist: Tay Ing Boon, who was attacked, did not have comparable strength of appearance and character, which may perhaps explain his vulnerability to attack.

Attacks and the threat of attacks did not deter either of the Tays from going deep into Iban territory. Both fondly reminisced about how easy it was to trade with the Ibans during their time. Tay Ing Boon remarked that one just needed to say how much an Iban's produce was worth, and he would accept it.[70] The traders probably went to longhouses where Ibans had little knowledge of the value of their jungle products or of the goods of the traders. According to both informants, times had subsequently changed and the Ibans no longer accepted a first price but haggled a lot instead over goods bought or sold.

A boat pedlar with a different experience from these two Tays was Lim Gu Piao, a Teochiu who came to Spaoh in the Batang Saribas, in 1923, at the age of twenty-nine.[71] When Lim first set foot at Spaoh, it had only six shops, three of which were owned by Hakkas. Lim had come to join a distant kinsman with whom he worked as a shop apprentice for about ten years, as was the case for most *sinkheh*. Only after his apprenticeship was over did Lim set out to try his fortunes as a boat hawker. Unlike most other traders who learned to speak the local languages, Lim did not and had to communicate in sign language. Lim said, 'I couldn't speak the Iban language and had to gesture and make signs with my hands.'[72] He stocked goods like food provisions, salt, tobacco, sugar and cloth, and his boat was capable of supporting a cargo of 7–8 *pikul*. Lim's dreams of returning to China were not fulfilled as he lost his savings in gambling, a common pastime for most boat hawkers in Sarawak. Boat hawkers who made the trip down-river once a month to a government station to renew their licences would take the opportunity to gamble.[73] When interviewed, Lim Gu Piao was in retirement, living with his children and grandchildren in a house, typically rural and *attap*-thatched with a hard mud floor.

Besides gambling, which was detrimental to the boat hawkers because of cheating and the lending out of money at exorbitant rates by the gambling-farmer who held exclusive gambling rights, the only other forms of recreational activity catered to in the bazaars were opium smoking and drinking (alcohol). Gambling and opium dens and saloons were operated on a monopoly basis, and these monopolies brought in valuable revenue for the Raj state and the monopoly farmer who held the franchise. The farmers were usually prominent *towkay*, and invariably acted as community leaders, for wealth among the *émigrés* was an index of social status. For those Chinese who were not married—whether to local women or wives brought over from China—drinking, smoking, and gambling were almost the only ways in which they could spend their leisure time and money, enjoy some degree of social life among members of their own speech group, and meet other Chinese, Malays, or Ibans. Brooke officials commented unfavourably on the influence of recreational activities like arrack drinking and gambling on the Ibans and the Malays. It would appear that the officials preferred the natives to keep away from these forms of entertainment, allowing them to be monopolized by the Chinese.[74] For boat hawkers who often travelled up-river for long periods of time, with no community life whatsoever, an occasional trip to the bazaar would find them frittering away their hard-earned money in gambling.

The monotony of bazaar life was relieved by the occasional visits of *wayang* (theatre) groups. A Chinese marionette company from Singapore arrived at Simanggang in 1898 to give a month-long performance. The visit was subsidized by Simanggang's leading merchants.[75] In April 1910, a Chinese *wayang* group of sixty performers arrived at Simanggang from Kuching, and it attracted wide interest 'for the bazaar was crowded with Chinese and Ibans from all over the district'.[76] A similar *wayang* performance in 1913 also attracted large crowds of Ibans.[77] These *wayang* shows gave a sense of ethnic identity to the Chinese, who were only a minority in an Iban district.

The cultural hiatus experienced by the early pioneers was accentuated by the lack of Chinese schools. The first Chinese school in the entire river basin was opened only in August 1911,[78] although the Chinese bazaars there had been in existence for at least half a century. Tan Khee Kiang still remembered this school when interviewed: in the 1920s, the school could boast of only eleven pupils.[79] The school taught only the barest rudiments of Mandarin. The *ulu* Chinese traders were too few in number to effectively remain a Chinese ethnic community. They were very much influenced by the natives, especially through intermarriage.

Although the new settlers generally had a liberal attitude to religion wherever they settled, they erected permanent structures for their deities. These deities provided spiritual and emotional solace for the Chinese in their new environment. The deity in Simanggang was the Goddess of Mercy, Kwan Yin, one of the most popular goddesses in the Chinese pantheon. It is not known when the temple in honour of the goddess Kwan Yin was first built in Simanggang, but it was rebuilt in 1899.[80] Pusa had its temple built in 1897,[81] and Roban in 1906.[82] Most of the bazaars, no matter how small they were, had a temple or a shrine.

The Rejang River Basin

Like those in the Batang Lupar district, the pioneering traders along the Rejang River were Hokkiens and Teochius. The Rejang Chinese traders had pioneering experiences similar to those of the traders in the Batang Lupar. They, too, had to face physical hazards such as floods and rapids, as well as personal danger in their dealings with the up-river Kayans and Kenyahs.

It is not known who the prominent *towkay* of Kanowit or Sibu were in the 1860s and 1870s, but by the beginning of the twentieth century, Teo Chong Loh, a Hokkien merchant based in Sibu, had made his way up the social and economic ladder in 'rags to riches' fashion. Teo arrived in 1897 from Singapore as a destitute fifteen-year-old boy, later owning shops in Sibu, Kanowit, and Kapit and becoming the spokesman and leader for the Hokkien community in the Rejang when he was appointed *kapitan*.[83]

Ordinary boat traders faced dangers like theft and murder, similar to dangers experienced in the Batang Lupar. Kayans and Kenyahs were less warlike and less nomadic, but they still indulged in head-hunting whenever the opportunity arose. Solitary itinerant traders provided excellent targets.

In 1894, a trader in Bintulu, Ah Liong, was killed at the mouth of the Penuan tributary of Belaga River by seven Kenyahs. Ah Liong had been living at the mouth of the Penuan River after moving from Bintulu. From the report of the incident, the exact motives for the murder cannot be deduced,[84] but the incident goes to illustrate that boat hawking was a risky occupation.

In the next incident, robbery was clearly the motive. In 1909, a trader, Tang Too Yong, was murdered in the Upper Rejang. Ibans had migrated to the Upper Rejang by the beginning of the century and were causing the Brooke administration concern, on account of their raiding attacks on their neighbours, the Kayans and Kenyahs. The trader fell victim to two Ibans, Janang and Isieng, who were later apprehended and brought before the court in Sibu:

Janang and Isieng were tried by jury at Sibu before a full court and were found guilty of the murder. From the evidence, robbery appears to have been the object of this dastardly crime. Tang Too Yong and another man were trading by boat and the two Ibans enticed the deceased into the jungle under the pretence of having *gutta* for sale, the other man, by name Sim Kai Cheng, remaining behind to look after the boat. By nightfall, as the deceased had not returned, Sim Kai Cheng became suspicious and armed himself while awaiting his friend, and later on, when the two Ibans returned and asked for permission to sleep in the boat, he refused them.... Sentence was deferred until the arrival of His Highness the Rajah, at Sibu on 21 June, when he presided in court. The Rajah, in summing up, described the murder of this inoffensive man as wanton and cruel, and stated further that much as the government disliked passing sentence of death on anyone, it was necessary in the interests of the peace and trade of the country. His Highness then passed sentence of death on Janang and Isieng.[85]

This report illustrated two points. First, the trade goods of the Chinese could provide temptation for robbery or murder. Second, the Brookes were ready to act in the interests of the traders. The next two oral accounts will show the relative degrees of safety and danger for traders down-river as compared with those up-river. Down-river traders were usually located in bigger bazaars and had the reassuring psychological comfort of Brooke forts located close by for protection. Up-river traders operated in more dangerous waters without such assurance, whether psychological or practical.

A down-river trader whose experiences seemed to be quite ordinary was Siaw Kee Siong, a Hokkien in Kanowit.[86] Born in 1900, he claimed that his father had been resident in Kanowit for over thirty years at the time of his birth. Siaw's father could have been among the pioneering group of Singapore Hokkiens who decided to settle and trade in Kanowit in the 1880s. In 1907, when Siaw was seven years old, he attended a makeshift Chinese school in a hut by the river bank, where there were twenty pupils altogether. They were taught by a teacher from China.

When Siaw was twenty years old, he became a trader, but he did not use a boat as was the norm. He carried his goods on his back. This form of peddling was more limited in scope: the trader was physically restricted by the load he could carry; it was more tiring; and the number of longhouses that could be visited was fewer. With the Ibans, the informant exchanged cloth, which was light to carry, for rice. This was a marginal form of trading and it took Siaw ten years to save up enough capital to own a boat with a burden of 10 *pikul*. He stocked cloth, salt, kerosene and biscuits, and food provisions for his travels. Unlike bazaars in the Batang Lupar basin, Kanowit's bazaar could boast of a social club. The Hua Siong Club, though a traders' association, was open to all; here, men would gather at night to tell stories of the past, and for gambling.

A pioneering family of the Upper Rejang was Neo Teck Hua's in Belaga.[87] Belaga-born Neo related how his father first reached Belaga in 1880 (four years before the establishment of a Brooke fort), as a young man of twenty. Neo's father was a boat trader who travelled in the uppermost reaches of the Rejang to barter goods with the Kayans, the Kenyahs, the Kajangs, and the nomadic Lesong Punans. He must have used a large boat for he had six native crew members. Above Belaga, the river branches out into the Batang Balui and the Batang Belaga, both with many sub-branches. The Batang Balui and Batang Belaga are beset by rapids and low water during the dry season, and boats have to be hauled. This must have been the reason why Neo had to have such a large crew. The stock of goods that Neo carried indicated a difference in commodities demanded by the Upper Rejang natives. The trader took along with him cloth, beads, and earthenware pots. Beads, especially old ones, had special prestige and ceremonial value for the Kayans and Kenyahs. The Upper Rejang was still a sparsely inhabited frontier region with exotic products like rhinoceros horns and bezoar stones which had extraordinary market value, in addition to the more ordinary *damar*, *gutta*, and *jelutong*.

Neo Teck Hua recalled an anecdote concerning his father having to flee from the Lesong Punans around 1910. Some Ibans, recent migrants to the Upper Rejang, had been raiding the Lesong Punans, and this could have motivated the latter to attack Neo's father and his party. Neo said, 'My father had a gun and this frightened off the attackers.'[88] As a result of this incident, Neo's father became a 'charismatic' figure with seemingly magical powers. This incident helped the trader to command awe and respect in his trading exchanges with the Punans. This anecdote parallels the experience (discussed in the previous section) of Simanggang boat trader Tay Bak Choon, who had to deal with the Skrang and Batang Ai Ibans, the much feared raiders and 'up-river rebels' of the Batang Lupar basin.

By 1937, Neo Teck Hua himself was a boat trader at the age of twenty-eight. He had a boat 50 feet long and weighing 5 to 6 *pikul*. Like his father, he needed a large crew of seven natives. Neo travelled between Belaga and Kapit, one of the most dangerous stretches of river to be crossed in Sarawak, covering a distance of 95 miles. When Neo had collected his products from the up-river Kayans in the Belaga district, he would journey down to Kapit and portage his goods across the 3-mile-long rock obstacles at the dangerous Pelagus rapids. It took two to three days to go down-river from Belaga to Kapit. The reverse trip was longer, taking anywhere from eight to thirty days, as the boatmen had to row against the current, a job made more difficult at low tide and with the low water-level of the dry season.

Pioneering life in the Rejang River basin was not much different from that in the Batang Lupar basin in terms of the difficulties and dangers faced by the traders. The trading bazaar network was dependent upon hardy boat pedlars who put their lives at risk by venturing far up-river.

The Baram River Basin

An invaluable source on pioneering life in the Baram district has been the Probate Books, which unfortunately are not available for other parts of Sarawak. These, together with the *Sarawak Gazette* and oral accounts, will help to reconstruct the lives of the first batch of permanent settlers in the Baram.

One of the first to open a shop in the Baram was Lim Goon Teng (also called Ah Hoon), who arrived in Marudi in the early 1880s. He was the owner of Chop Kim Wat and was to become one of the wealthiest and most influential *towkay* of Marudi. When he died in 1905, the year the *kapitan* system was started, Lim's son, Cheng Soon, was appointed the first *kapitan*.[89]

However, among the many hopeful Chinese pioneers who arrived in the Baram, Lim Goon Teng's material success was to prove the exception rather than the rule. As the following examples illustrate, trading in Sarawak's interior river basins was fraught with hazards, and for only a few fortunate traders was there the opportunity to enjoy the hard-earned fruits of their labour.

Only one year after the official founding of Marudi in 1883, Kim Sun, the sole Chinese trader in Niah, 60 miles from the mouth of the Baram, was found murdered in his own house, with his head gone.[90] No clear motives were established for this murder, but it highlights the point that traders put their lives at risk wherever they went. With no medical care in the outstations, solitary traders who fell ill could die. If they were boat hawkers in some isolated river or stream, they might have had no fellow clansmen upon whom to rely. Help, if it was forthcoming from people at the bazaars, would depend on whether the sick, injured, or deceased belonged to the same speech group and whether there were generous benefactors from among the more well-to-do *towkay* who were willing to help. Sick traders, when they had no one else to turn to, often sought the help of the local populace, as this case illustrates:

On 23 January 1890, the body of a Chinese trader was brought down by some Kayans who stated that the man had waged them to bring him down but that he had died on the way, and one of the shopkeepers received a letter by the deceased stating that he did not expect to reach here alive as he was ill with fever.[91]

Attacks similar to those which were suffered by Batang Lupar and Rejang River traders occurred in the Baram. In November 1894, Ah Pin, a Chinese trader at Long Labong, one of the uppermost reaches of the Tinjar tributary, was murdered in his boat at night.[92] That he was trading alone so far away from Marudi was remarkable. A single boat trip up-river from Marudi to Long Labong could take anything from two to four weeks, with rapids and low water to negotiate, although the return journey would be faster—possibly about a week. The stocks the boat hawker could carry would be limited. The number of longhouses he would be able to stop at along the way would be few, this being a sparsely populated region. The trader would have to depend entirely on his own rations. At best, boat hawking was a marginal occupation, and many did not accumulate the hoped-for wealth that had compelled them to come to Sarawak in the first place. The Baram Probate Books reveal this material aspect of the pioneering life of the boat traders and bazaar *towkay*.

A perusal of the Probate Books shows statements of the traders' assets and liabilities. These statements, recorded after their deaths, point to characteristics of the credit structure, the types of exchange goods used in barter trade, the monetary returns from conducting trade in the Baram and, lastly, the network of social relations.

The following is an example of the assets of a successful Teochiu shopkeeper in Marudi, Chap Long, who died in 1889, leaving behind considerable assets.[93] He had an Iban wife, Ah Moi, who took over the business, which was unusual. Chap Long had a mixed clientele

Chap Long's Estate and Assets

Value of goods in stock	$1,200.00
22 pigs valued at	200.00
1 garden	100.00
7 shops	$2,800.00
Credit:	
Malays	$ 300.00
Dayaks	$1,400.00
'Belait people'	$1,000.00
Kayans	200.00
	$7,200.00

of local Malays, Ibans, 'Belait people' (Brunei Malays), and Kayans. Less than half of his assets consisted of advances to native clients. In an economy where the use and availability of money was limited, there was a prevalence of this type of long-term 'exchange' whereby the goods of the traders were advanced on the promise of future payments in kind—that is, in jungle products. Success in trading in rural Sarawak did not depend solely on how much was bought or sold, but also on the willingness to take risks in advancing goods, and on how much of these advances would be recoverable. Most if not all traders were cognizant of the fact that a certain percentage of debts would always be unrecoverable due to clients absconding, or because of practical difficulties in approaching debtors. Estimates of unrecoverable debts range from 30 to 40 per cent of total advances, according to elderly Chinese informants.[94]

The trader himself was dependent upon the goodwill of Kuching *towkay* and the amount of credit he was able to obtain from them; the Kuching creditors in turn were indebted to Singapore traders. The whole trading structure hinged on credit, stretching all the way from the native client to Singapore suppliers.

Another source of 'working capital' for the *ulu* trader was 'Dayak deposits'. The Ibans, as missionaries and officials had observed, were not given to frivolous spending, but were frugal, depositing hard-earned cash with shopkeepers, or else converting it into the purchase of old ceramics which had a lasting value.[95] As will be seen from the list of Chap Long's debts, below, he owed money to Kuching shopkeepers as well as to Ibans, acting as some kind of informal 'savings bank' for the latter.

Chap Long's Debts

To Kuching *towkay*:	Kwong Wat	$ 308.08
	Tek Ann	169.92
	Soon Hap	267.73
	Others	1,104.27
		$1,850.00
'Dayak deposits'		$1,300.00
Baram *towkay*		150.00

Another example from the Probate Books—the probate statement on Chin Poh, a Hokkien, who died in 1900[96]—revealed the extent of Kuching's control over *ulu* trade, not only in granting credit, but in actual ownership of business as well. The probate statement showed that Chin Poh was a partner in the firm, Seng Ann, owning seven shares out of a total of thirteen. The business of Seng Ann was owned by Yong Ann and Company of Kuching. The credit and assets statement bears a similarity to the previous example, with a substantial portion of trade—in this case, roughly half of it—predicated upon advances owed by others, such as Ibans, Kayans, and possibly Chinese boat hawkers.

Chin Poh's Credit and Assets

Credit:	
Ibans	$ 659.20
Kayans	465.30
Chinese	2,658.59
Traders	1,306.59
Goods in shop (e.g. *gutta*)	593.61
Goods in shop (e.g. cloth)	4,002.15
	$9,685.44
Debit:	
'Dayak deposits'	$ 350.50
Chinese deposits	162.23
Yong Ann and Co., Kuching	7,877.30
Hong Tung, Kuching	676.84
Tong Seng, Kuching	13.65
	$9,080.52

After the assets—that is, the goods in the shop—had been apportioned among the surviving partners, the deceased was still in debt to the firm of Yong Ann, owing $203.75. All Chin Poh had, after many years of trading, were savings of $140 left to his daughter; this amount was insufficient to meet his outstanding debts. This example fits the classic illustration of traders who died as paupers. There were many more like him. Being a shopkeeper, the deceased was at least saved the physical and financial hardship faced by those engaged in up-river trade.

Most boat pedlars operated with the minimum of financial capital. There was the case of Vi Ah Cham, a Hokkien trading in the Tinjar tributary area, who died in 1901 'with absolutely no property ... a hopeless case, and no recovery (of assets) can be expected'.[97] He was indebted to Marudi shopkeepers in the following amounts: Soon Kiat $6.54, Joo Watt $12.13, Ah Wa $2.60, Seng Ann $10.99, Teong Kiat $47.44, and Raman $4.00. The small sums owed by Vi Ah Cham probably meant that he was trading on a limited scale.

Some other up-river traders were in a more fortunate position than Vi Ah Cham. Ah Hiong of Long Sungei on the Tinjar River left behind some property when he died in 1915.[98] The probate statement on Ah Hiong contained a list of his belongings. This is a particularly important list for it shows the type of goods given and received in barter trade by traders in the Baram at that time, along with their worldly possessions.

Ah Hiong's Assets

Rattans, 205 *gelong* valued at $81.16
4 *parang* (long knives)
4 spears
5 boxes
2 brass trays
5 plates
2 cups
1 blowpipe
5 mats
5 singlets
6 *chelapa* (oblong box for keeping accessories)
1 bundle cigarette papers
4 packets of matches
1 *labong* (forge iron)
1 *datching* (weighing scale)

1 book of Chinese characters
1 *jala* (net) chain
1 pair of spectacles
2 bottles of perfumery
1 plaited mat
3 ties of beads
1 small rattan basket
1 *chawat* (loin-cloth)
1 sack of *sut* (outer clothing)

The goods listed reflect a diversity of items, from prestige goods such as brass trays and beads to utilitarian goods in the form of plates, cups, and matches, to clothing and decorative trivia such as perfumes. This particular trader had goods which appealed to the varied tastes of the up-river natives.

Ah Wa, who died in 1915, was a pedlar at Long Tiram, and he had an even shorter list of goods.[99]

Ah Wa's Assets

12 tins of Kampar tobacco
1 tin of brass coil
1 *tawak* (gong)
2 gongs
2 *chanangs* (shallow gongs)
2 paddles
1 piece of *blachu* (black cotton cloth)
1 red cloth
3 *chita* (cotton chintz)
3 sarongs

Ah Wa had light items such as cloth and tobacco, and prestige goods like gongs, all of which were unperishable and therefore useful for long river journeys.

A somewhat different inventory showing an absence of prestige and utilitarian goods, but including a quantity of foodstuffs, was recorded in the probate statement on boat hawker Ah Wai who died in 1901.[100] He died in possession of one boat, 1 *kati* and 10 *tahil* of tobacco, 60 *kati* of salt, and 16½ *kati* of sugar. There appeared to be some kind of specialization among up-river traders, so that there were those who peddled in luxury and prestige items and those who carried only foodstuffs.

The Probate Books occasionally revealed whether the traders had wives in China or not, and also whether or not they were married to local women. When Lee Ah King died in 1915, he left behind a local wife, an 'Orang Bukit' woman (presumably a Kelabit) whom he had married by civil law. Previous to that, he had been married to a 'Belait' (Brunei Malay) who had died before his subsequent marriage. In addition, he had a Chinese wife in Fukien province in China.[101] Another up-river hawker, Be Ah Wai, trading in the Temadoh River area, was married to a Sebop woman when he passed away.[102] Oh Kee, a trader on the Tutoh River, who died in 1906, was married to a Long Patah Kayan woman. He was reported to have a wife and two sons in China.[103] These examples illustrate a more general pattern of marital unions between migrant traders and indigenous women; these were not uncommon in rural Sarawak.

The Probate Books have provided valuable information on pioneering life along the Baram River—information such as the credit system upon which trading operations depended, material evidence of economic exchange, and the extent of social relations. The information was, however, recorded at a particular time—that is, after the deaths of the traders. As such, it is a static presentation and does not inform us of the dynamics of trading activity, nor of the social processes and practical difficulties underlying economic exchange. To highlight some of the more difficult pioneering conditions of *ulu* trade, such as those faced in river communication and competition with Brunei Malay traders, some oral accounts will be reconstructed.

An example of a boat hawker is Tien Chu Kim, a Teochiu born in Marudi in 1905, whose father arrived at the Baram frontier in 1890, eight years after a permanent settlement was established there.[104] Tien recalled a story related to him in his childhood by his father. The father was a boat hawker at a time when trading competition between Chinese and Brunei Malays was still keen. A group of Brunei Malay traders, in debt to the father, quarrelled with him. One day, armed with poles, they set upon his boat and surrounded it with the intention of attacking the trader. He challenged the group of twenty persons to fight with him, asking them to replace their poles with slabs of iron from his boat. The boat trader then said, 'Those of you who want to die should come forward.' The would-be assailants then quietened down and suddenly left, having lost their courage.

This episode is similar to those recounted earlier from the Batang Lupar and Rejang River basins. These anecdotes show how dangerous boat hawking was as a livelihood. Such dangers could be circumvented by blood-brother pacts between the Chinese trader and the native, but there is little evidence that this was a common practice.[105] The rural traders were a tough and hardy breed, who sometimes lost their lives at the hands of the local populace despite their bravery.

In 1919, at the age of fourteen, Tien Chu Kim followed his father in boat hawking. He rowed his boat from Marudi to Long Lama, a journey taking three to four days, sometimes a week. The products from Long Lama were bird's nest, rice, and *gutta*. As exchange items, Tien carried goods such as sugar, salt, kerosene, white wine, cloth, tobacco, and canned food. A pass, not a hawker's licence, was needed to go up to Long Lama. Tien remarked that it was easy to gain the upper hand over the Kayans and Kenyahs in trading transactions. However, the indigenous people were not to be underestimated as they retaliated by simply absconding from their debts.

A boat hawker who had to risk his life regularly crossing dangerous rapids was Yap Poh Chai.[106] Yap, a Hokkien, was born in Marudi, son of a Long Akah trader. He did not receive any formal education and, in 1936, turned to boat hawking at the age of sixteen. He took his boat from Long Akah to Lio Matu, 60 miles away and one of the last few settlements in the uppermost reaches of the Baram River, and to Long Lellang, 35 miles away in the Akah tributary.

Yap used a sizeable boat and five native helpers were needed to pole it. The boat weighed 50 *pikul*, and was 36 feet long and 3 to 4 feet wide. These dimensions made it easier for Yap and his party to pass through hazardous rapids on the way to Lio Matu. The trader and his companions could make a safe journey across the rapids only when the water was shallow. Water-levels could rise suddenly after a torrential downpour, and would then drop almost as rapidly when the rain had eased. Sometimes the boatmen had to wait for days at the foot of rapids for the right water-level before proceeding further.

Whenever Yap felt he had collected enough produce from the Upper Baram, such as *damar*, *gutta*, and bezoar stones, he would travel down to Long Lama and straight on to Marudi. A journey from Long Akah to Long Lama might take one to two days, and from Long Lama to Marudi up to three days. Yap would make these trips down to Marudi about twice a year.

These oral accounts have highlighted the experiences of boat hawkers. The river was a source of livelihood for the boatmen who, despite their land-based background, readily took to the water, thereby demonstrating their adaptive qualities in an alien environment. Boat traders played an important role in the expansive riverine trading networks, bringing goods right up to the longhouses and collecting jungle products from the natives. Without the sweat and toil of the boat pedlars, trade might not have expanded as rapidly

THE ORIGINS AND PIONEERS OF *ULU* TRADE 97

and extensively as it did in *ulu* riverine Sarawak in the nineteenth and early twentieth centuries.

1. Rivers here are used interchangeably with Divisions: the Batang Lupar refers to the Sri Aman Division; the Rejang River means the Sibu and Kapit Divisions; and the Baram River, the Miri Division.
2. Benedict Sandin, *The Sea Dayaks of Borneo before White Rajah Rule* (London: Macmillan, 1967), p. 64.
3. Ong Ngee Guan, 'Teochew Chinese in Simanggang', in Teochew Association, Kuching, *Centenary Volume* (Kuching: Teochew Association, 1965), p. 102 (text in Chinese).
4. *SG*, 1 April 1872, p. 27.
5. A. B. Ward, *Rajah's Servant*, Cornell University Southeast Asia Program Data Paper No. 61 (Ithaca, New York: Cornell University Press, 1966), p. 27.
6. Ibid.
7. *SG*, 1 August 1883, p. 75.
8. *SG*, 1 February 1872, p. 13.
9. Interview with Bong Nam Siong, 10 April 1981, Engkilili.
10. Robert Pringle, *Rajahs and Rebels* (London: Macmillan, 1970), p. 56.
11. *SG*, 30 March 1871, p. 54.
12. Interviews with Sim Ah Lah, 8 April 1981, and Lim Chen Tau, 9 April 1981, Betong.
13. Pringle, op. cit., p. 205.
14. *SG*, 1 February 1916, p. 35. In the oral traditions of the Spaoh Chinese, it is said that a boat hawker requested the Resident for permission to set up shop on land. See Chen Khoon Yan, 'A Brief History of the Opening of Spaoh', in Teochew Association, Kuching, *Centenary Volume* (Kuching: Teochew Association, 1965), p. 107 (text in Chinese).
15. *SG*, 1 August 1885, p. 70.
16. Allied Geography Section, Southeast Pacific Section, *Area Study of Sarawak and Brunei*, Vol. 1, September 1944, p. 168.
17. *SG*, 1 March 1890, p. 34.
18. R. Outram, 'The Chinese', in Tom Harrisson (ed.), *The Peoples of Sarawak* (Kuching: Government Printing Office, 1959), p. 119.
19. Frederick Boyle, *Adventures among the Dayaks of Borneo* (London: Hurst and Blackett, 1865), p. 98.
20. Interview with Syn Chin Joo, 29 June 1981, Kapit.
21. Interviews with Siaw Kee Siong, Lim Zean Khuan, and Lim Choon Jin, 26-28 June 1981, Kanowit.
22. *SG*, 24 January 1871, p. 38.
23. Peter Stevens, 'A History of Kanowit District', Part 2, *SG*, 28 February 1971, p. 34.
24. *SG*, 2 August 1875, p. 6.
25. *SG*, 16 January 1877, p. 2.
26. *SG*, 22 January 1878, p. 3.
27. Pringle, op. cit., p. 255.
28. *SG*, 1 March 1888, p. 33.
29. *SG*, 1 July 1887, p. 123.

30. Ibid.
31. *SG*, 1 August 1887, p. 133.
32. *SG*, 1 February 1893, p. 29.
33. *SG*, 16 July 1872, p. 54.
34. Cited by Richard Goldman, 'The Beginnings of Commercial Development in the Baram and Marudi', *SG*, 31 March 1968, p. 54.
35. In July 1882, Rajah Charles Brooke obtained the Baram concession from the Sultan of Brunei. Marudi's original name was Claudetown, named after Claude Champion de Crespigny, the first Resident of the Division. *SG*, 1 July 1882, p. 34.
36. *SG*, 1 May 1883, p. 50.
37. Ibid.
38. Ibid.
39. *SG*, 1 August 1885, p. 72.
40. The Rajah's Order Book, 12 April 1889, p. 425, SA.
41. *SG*, 1 June 1889, p. 89.
42. *SG*, 1 April 1891, p. 56.
43. Goldman, op. cit., p. 60.
44. *SG*, 1 August 1884, p. 87.
45. Cited by Goldman, op. cit., p. 60.
46. *SG*, 1 June 1888, p. 78.
47. *SG*, 1 June 1891, p. 78.
48. *SG*, 3 March 1906, p. 55.
49. Interviews with Kang Chiat Sam and Lee Chin Tek, 10 August 1981, Long Lama.
50. *SG*, 16 September 1919, p. 247.
51. See note 3.
52. *SG*, 1 December 1899, p. 353.
53. Ibid.
54. A. B. Ward, op. cit., p. 78.
55. *SG*, 1 March 1889, p. 42; 1 August 1894, p. 130; 7 June 1901, p. 135; and 2 July 1903, p. 136.
56. Interview with Madam Tay Lan Eng, 30 March 1981, Simanggang.
57. *SG*, 5 June 1907, p. 150.
58. *SG*, 1 October 1897, p. 173.
59. *SG*, 1 April 1913, p. 72.
60. *SG*, 1 April 1895, p. 70.
61. *SG*, 3 October 1906, p. 241.
62. *SG*, 1 February 1910, p. 35.
63. Monthly Report, 1 June 1871, p. 250, SA.
64. *SG*, 16 April 1914, p. 96.
65. *SG*, 1 June 1929, p. 100.
66. Interviews with Madam Tay Lan Eng, 30 March 1981, Simanggang, and Lim Thai King, 22 April 1981, Lubok Antu. For more details on the up-river 'rebels', see Pringle, op. cit., Chapter 8.
67. A number of traders with whom the author talked said that to maintain good rapport with their Iban clients, they did not force the Ibans to repay their debts if they were unable to, or refused to do so. The presentation of gifts by the traders to the natives was an act of goodwill to reciprocate the hospitality extended by the longhouse-dwellers, such as overnight accommodation. Interviews in Simanggang with Tay Ing Boon, 27 March 1981; Yeo Yiaw Piaw, 27 March 1981; Tay Bak Choon, 2 April 1981; and Lim Gu Piao, 9 April 1981; in Roban with Ng Seng Phua, 14 April

1981; and in Saratok with Ong Tiang Boon, 13 April 1981.
68. Interview with Tay Bak Choon, 2 April 1981, Simanggang.
69. Interview with Tay Ing Boon, 27 March 1981, Simanggang.
70. Ibid.
71. Interview with Lim Gu Piao, 9 April 1981, Simanggang.
72. Ibid.
73. Information from interview with Ong Tiang Boon, 13 April 1981, Saratok.
74. *SG*, 4 January 1897, p. 17; and 1 February 1913, p. 33.
75. *SG*, 1 July 1898, p. 240.
76. *SG*, 16 May 1910, p. 109.
77. *SG*, 2 June 1913, p. 118.
78. H.H. The Rajah Confidential, Vol. 4, 20 August 1911, SA.
79. Personal information from Tan Khee Kiang, 1 April 1981, Simanggang.
80. *SG*, 1 July 1899, p. 240.
81. *SG*, 4 January 1897, p. 17.
82. *SG*, 3 March 1906, p. 71.
83. Outram, op. cit., p. 120.
84. *SG*, 1 December 1894, p. 204.
85. *SG*, 1 July 1910, p. 139.
86. Interview with Siaw Kee Siong, 26 June 1981, Kanowit.
87. Interview with Neo Teck Hua, 2 July 1981, Belaga.
88. Ibid.
89. Information obtained from Lim Poh Kui, son of Lim Cheng Choon, 11 August 1981, Marudi.
90. *SG*, 1 March 1883, p. 26.
91. *SG*, 1 April 1890, p. 53.
92. *SG*, 2 January 1895, p. 15.
93. 14 September 1899, Baram Probate Book (hereafter BPB), Vol. 1, p. 3, SA.
94. See note 67.
95. Edwin Gomes, *Seventeen Years among the Sea Dayaks of Borneo* (London: Seely and Co., 1911), p. 63.
96. 30 November 1900, BPB, Vol. 1, p. 12, SA.
97. 31 July 1901, BPB, Vol. 1, p. 12, SA.
98. 7 April 1915, BPB, Vol. 1, p. 270, SA.
99. 19 July 1915, BPB, Vol. 1, p. 275, SA.
100. 25 June 1901, BPB, Vol. 1, p. 18, SA.
101. 22 June 1915, BPB, Vol. 1, p. 272, SA.
102. 25 June 1901, BPB, Vol. 1, p. 18, SA.
103. 10 August 1906, BPB, Vol. 1, p. 154, SA.
104. Interview with Tien Chu Kim, 8 August 1981, Marudi.
105. In the course of the author's interviewing, he came across only one instance of a claim of a blood-brother pact concluded between a Kayan chief and a Chinese boat hawker in the Tinjar River. After the symbolic adoption of the trader as the chief's son, the boatman was free to travel along the Tinjar River. Information obtained in an interview with Kang Chiat Sam, 10 August 1981, Long Lama. From written sources, there is an example of a brotherhood pact involving Robert Burns, a grandson of the famous Scottish poet, one of the first Europeans to visit the Kayans of the Rejang River in 1847. Burns made a brotherhood pact with Lasa Kulan, a Kayan chief in Belaga. See Robert Burns, 'The Kayans of the Northwest of Borneo', *JIAEA*, Vol. 3 (1849), pp. 146–7.
106. Interview with Yap Poh Chai, 12 August 1981, Long Lama.

5
Organization of *Ulu* Trade: Economic Exchange and Transformation

THE preceding chapter was concerned with the growth of bazaars and the pioneering lives of the Chinese traders. It also dealt with the traders' efforts in encouraging their indigenous clients to trade. This chapter will look at the way in which economic exchange was organized, and at the subsequent transformation of the riverine economies with the active participation of the natives in the process of exchange. The effects of trade on indigenous society will be considered, as will the nature of barter trade, the use of money as a medium of exchange, and the workings of the credit system utilized in Sarawak (graphically illustrated in Figure 1). The similarities and differences between *ulu* trade in the three riverine basins will be discussed.

Products, Procurement, and Exchange

The tropical jungle of Borneo contains many varieties of trees and creepers which yield useful by-products such as fruits and gums. A number of animals and birds inhabit the forest and they can be hunted for game and for certain valuable portions of their anatomy.

Among the better-known gums which can be extracted from more than twenty species of trees is *gutta percha*.[1] It is used locally for lighting torches, for caulking and repairing boats, and as a domestic adhesive. In the 1870s and 1880s, there was an industrial demand for this product in the West, where it was used as insulation for cables. *Damar* is another type of resin used for caulking boats, but in nineteenth-century Europe and America, it was used as an ingredient in electric cables and as marine glue. Beeswax, another product of the jungle, is collected from combs built by wild bees in the tall branches of the big *tapang* trees. The riverine vegetation and thick forest cover of Borneo favour the growth of certain tree creepers, one such variety with economic value being the rattan creeper. Rattan creepers are stems of climbing plants which grow entwined around tall, unbranched trees.[2] The rattans grow best in well-drained soils and along the edges of streams and fresh-water swamps. Rattan is

Figure 1   The Domestic Impact of *Ulu* Trade

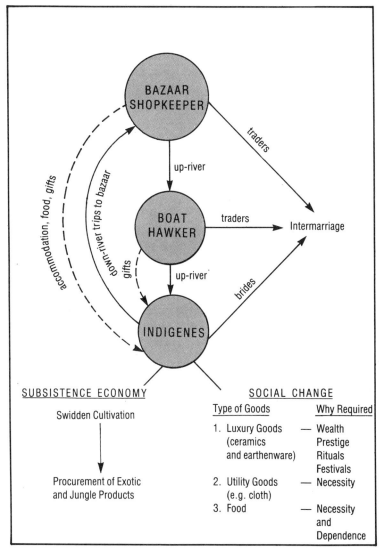

used domestically for binding and cordage, but in the nineteenth century there was a demand for it in Singapore, China, Europe, and America, where it was needed for the furniture and basketry industries. Rattan is abundant in the Upper Rejang and Upper Baram River basins.

A tree type common in Borneo is the *Shorea isoptera* species which yields oily fruits. This tree, which grows well by the river banks and can be planted, is locally called *engkabang* or illipenut, and is named after the tallow-like substance extracted from its fruit. It is used for cooking and as food by the nomadic Penans, but the fat of the fruits has other uses as well in the candle and soap-making industries. The fruits, which are seasonal, are a boon to the indigenes.[3] *Engkabang* was sought after by the Chinese traders in the nineteenth century as an export commodity.

In the nineteenth century, there were other products which found a ready market in Singapore and China. Camphor, used in Chinese medicine, is found in the trunk of a certain species of tree. Exotic products found in Borneo upon which the Chinese placed a high value for their medicinal and reputedly aphrodisiac properties were rhinoceros horns and bezoar stones, the latter being the intestinal and gallstone secretions of a particular type of monkey. A luxury food item was bird's nest, made from the salivary secretions of swiftlets in caves.

The three river basins were differently endowed with jungle and exotic products. The Batang Lupar and its two adjacent rivers in the Second Division, the Saribas and the Krian, had *gutta percha*, *damar*, and rattan. The Batang Lupar basin was not a rich region and was dependent on rice as an exchange commodity and on the seasonal *engkabang*. The Rejang and Baram basins, being abundant in *gutta*, rattan, and exotic products, were not dependent upon rice for exchange, nor on *engkabang*.

A constant problem for the Batang Lupar traders was that of enticing the natives to procure the jungle products. The traders' reliance on the Ibans for this purpose was never more apparent than in the fall-off in trade when cholera struck, as it did periodically. In 1877, when there was a cholera outbreak, 'trade was somewhat depressed on account of the cholera having frightened all the Ibans away'.[4] A similar epidemic in 1888 caused trade to slacken and 'there was a general absence of cargo for the *bandongs*'.[5]

The Batang Lupar exported rice, padi, *gutta*, and *jelutong* in that order of importance.[6] The ripening of *engkabang* fruits provided an occasional windfall for the native collectors. In February 1912, nearly 2,500 *pikul* of the nuts were exported to Kuching from the Batang Lupar basin and, at the beginning of April of the same year, there were another 1,200 *pikul* awaiting shipment.[7] There was another *engkabang* boom in 1914:

It is unfortunate that just at a time when there is a boom in a trading commodity, money should be scarce in the bazaar. The depression in Kuching had made itself felt throughout the district and although there are big profits to be made for the Kuching merchants and their middlemen in the outstation bazaar through illipenuts, yet the latter complain they cannot obtain sufficient cash advances to purchase with. Children are disposing of their properties, and women of their jewelleries to provide means, and wherever possible, the traders persuade the Ibans to take brassware and goods in exchange for the nuts.[8]

The Simanggang credit structure faltered in the face of an unexpected harvesting of the illipenuts. As a windfall like this came only periodically, it was necessary for the Simanggang traders to react rapidly and to find means to pay off their Iban clients. Despite the lack of sufficient credit, the *engkabang* boom continued unabated until 1915, resulting in Simanggang 'becoming deserted as everyone was away harvesting or gathering *engkabang* fruit'.[9] Four years later, another *engkabang* crop was reported: 'A great number of *Ibans* would not sell the *engkabang* they have for notes, and in consequences are holding it up until they can get silver. There is little copper to be had now, and although I (F. A. W. Page-Turner) have offered to exchange silver for copper, nobody has taken advantage of the offer.'[10] There was a preference for silver or copper coins which could be used as ornaments. Promissory notes tended to be treated with suspicion.

While trade in the Batang Lupar depended on the sale of rice and the procurement of *engkabang*, and these could be affected by epidemics amongst the indigenous population, the Rejang River basin was more fortunate. It was a frontier region rich and alluring, with a plentitude of *damar*, *gutta*, rattan, beeswax, timber,[11] camphor,[12] rhinoceros horns, and bezoar stones.[13]

Rattan was an especially important export from the Rejang. In a single month (that of January 1886), the Chinese in Sibu had 1,818 *pikul* of rattan ready for loading.[14] During June of the same year, when the price was $35 per 100 canes, schooners shipped directly to Singapore 1,898 *pikul* of *rattan sega*, 102 *pikul* of *tot* (rattan bud grafts), and 25,000 Malacca canes.[15] Not all of the Rejang trade went directly to Singapore. Part of it was diverted through Kuching.[16] Five years later, in 1891, the Rejang River bazaars were said to be '... full of *rattan*, with the traders anxious about exporting them while the present high prices prevail. In three weeks' time, they will have 46,000 bundles of *rattan* ready for export, 4 times as much as the *Adeh* can carry. The *Ibans* are getting ready $60–80 for a 100 cane-bundle.'[17]

The rattan came mostly from the upper reaches of the Rejang River and immediately below it.[18] Collecting rattan was a labour-intensive effort for the Kayans, Kenyahs, and Ibans, who had to cut the canes and send them by river down to the bazaars. The cane was collected by cutting through the stem of the creepers near their roots. It was then hauled through the jungle to the river banks where the stems, originally 150–200 feet long, were cut into lengths of about 15 feet and left to dry in the sun. When the canes were dry, a large number of bundles would be bound together to form rafts. On each raft a hut would be erected, and two or three men would navigate the raft down-river to the nearest bazaar.[19] In the Upper Rejang River basin, the nearest bazaars were Belaga and Kapit. From Kapit, small ocean-going vessels and Chinese junks would take the rattan down to Sibu.

Rattan trade in the Rejang was as unpredictable as the *engkabang* windfall in the Batang Lupar. A glut in the rattan market would cause prices to drop. In 1909, for example, 'there was a large quantity of rattans and other jungle produce in the Belaga bazaar in spite of the low state of the market'.[20]

While the Batang Lupar was troubled by epidemics, the bottleneck to trade in the Rejang was the lack of communication with Singapore and Kuching. In February 1881, during the wet monsoon, no vessels were available for loading and transporting the products stored in the Sibu bazaar.[21] A group of nine Sibu *towkay* offered to guarantee 40 *pikul* of *gutta* and 11,600 bundles of rattan as cargo for one shipment for the Singapore and Sarawak Steamship Company vessel, the *Adeh*, if she were prepared to sail to Singapore. These shipping difficulties increased in later years, as was reported in 1919:

A serious difficulty has arisen owing to the shipping facilities being absolutely inadequate to cope with the increase in cargoes for export. The Chinese have made every effort to charter a steamer to Singapore, but without success. All the bazaars are choked with illipenuts and it is estimated that only a quarter of the nuts have as yet been put on the local market.[22]

The bazaars of the Rejang River basin, stretching from Sibu to Belaga, were usually well stacked with rattan and *gutta*, in stark contrast to the normally empty bazaars in the Batang Lupar basin.

A frontier region similar to the Rejang was the Baram River basin. Like the Rejang, the Baram had one commodity—rattan—in plentiful supply, in addition to *gutta*, rhinoceros horns, and bezoar stones.

A fort had been established at Marudi on the Baram River in 1883, and a number of *attap*-houses were built to serve as the bazaar. A year later, the value of Marudi's exports had risen to $62,655,[23] and the government office was collecting export and import duties. By the first quarter of 1884, the value of exports had risen to $68,584.[24] Marudi was served by the Singapore and Sarawak Steamship Company vessel, the *Adeh*.[25] The *Adeh*'s itinerary included the bazaars of the Rejang. In a single trip in March 1886, after the deluge of the end-of-the-year monsoon was over, the *Adeh* carried away a cargo of 9,000 bundles of rattan in addition to *gutta*, camphor, and bird's nest. Another trip in February 1887 saw the *Adeh* take away 10,000 bundles of *rattan gemaing*, 2,000 bundles of *rattan sega*, over 120 *pikul* of *gutta*, and half a *pikul* each of camphor and white bird's nest.[26] In addition, more than 40,000 bundles of rattan were reported to be ready for shipment, and rattan rafts were arriving from up-river daily.[27]

The increasing volume of rattan in Marudi was making it difficult, if not impossible, for the *Adeh* to cope single-handed. A *Sarawak Gazette* report in May 1887 stated:

Rattans are coming down in increasing quantities and during last month, about 2,500 bundles of *sega* and 8,000 bundles of *gemaing* arrived, chiefly from the Tinjar. The *Adeh* left taking a cargo of 1,280 bundles of *sega* and 5,000 of *gemaing*. Rattans have been coming down steadily throughout the month, and there are now in the bazaar 22,000 bundles of *gemaing* and 2,700 bundles of *sega*.[28]

The magnitude of trade was such that the bazaar Chinese could not clear their stocks of *gutta* and rattan due to shipping uncertainties. Again, ten years later in February 1897, the Chinese shopkeepers stated that they had over 500 *pikul* of *gutta* valued at $40,000 and about 2,600 *pikul* of rattan awaiting shipment: 'The traders were obliged to store their *rattans* in large blocks covered up with *kajangs* on the road in front of the bazaar, as they had no more room for them in their warehouses.'[29] In August of the same year, 30,000 bundles of rattan and 200 *pikul* of *gutta* were ready to be sent away. The *Sarawak Gazette* reported that 'the Chinese will be glad to see the *Adeh* again as their "godowns" are quite filled up with jungle produce'.[30]

Although quantities of rattan were constantly being brought down to Marudi, the vagaries of trade were similar to those experienced in the Rejang and Lupar river basins, with fluctuating prices for the jungle products. During the first few years of the twentieth century, poor trade was recorded for the Baram district. The Baram Annual Report for 1904 stated that 'trade was in a very depressed state owing to the low prices obtainable for *gutta percha* and *rattan sega*'.[31] But the situation had improved somewhat by 1909. The Singapore and Sarawak Steamship Company steamers, *Kaka* and *Gladys*, calling once a month during the dry season, were unable to meet the demands of the Baram merchants.[32] When the *Gladys* arrived in March 1910, 'she was badly wanted as all foodstuffs except rice had run out in the bazaar'.[33] When the steamer departed, she took out a consignment of rubber and *jelutong* valued at over $10,000.[34]

Shipping difficulties which troubled the Rejang traders plagued the Baram merchants as well. An insufficient number of vessels going to Marudi was not the only reason. There were practical reasons which prevented ships entering the Baram River. A major hindrance to trade and transport between Kuching and Marudi was the bar at Kuala Baram, almost impassable during the rainy *landas* season from October to March. During this time, trade was at a standstill because of a lack of goods from outside to exchange for products of the Baram. Even when more ships called at Marudi after 1910, there were complaints from the traders of insufficient transport, complaints which were supported by the *Sarawak Gazette*. There was considerable dissatisfaction with the service provided by the Singapore and Sarawak Steamship Company.

A *Sarawak Gazette* report of April 1912 commented:

Great inconvenience is experienced and many are the complaints constantly made by the Baram traders owing to the want of sufficient and regular communication with that river. Baram is closed from October to March, by which time provisions are generally at a low ebb. The Steamship Company is a money-making concern and goes in the first place where it can find cargo, but by the terms of their agreement made with the government at the time they took over the coasting trade, the company cannot refuse any cargo offered for the Baram during the fine weather, and we should have thought that, after being cut off for five or six months, there could have been no difficulty in finding a full cargo for Baram and sufficient for a return journey to justify sending a steamer there specifically. Truly, there were complaints when the coasting steamers were run by the government, but they are worse now. Are the traders without whom the company would not exist, entitled to some consideration in the matter?[35]

Marudi's distance from Kuching, the bar at the mouth of the Baram, and the unreliability of commercial shipping services were factors which presented problems for the growth of trade in Marudi. Further, the monsoon season from October to March hindered shipping and resulted in the further isolation of Marudi bazaar. Trading junks from Hainan island, calling at Marudi at the end of the monsoon season, helped to break the isolation: 'The junks would arrive laden with salt and deck cargoes of fowls, ducks, pigs and other livestock. For weeks, they would lie with characteristic patience inside the bar, until a favourable wind would enable them to make the ascent to Marudi, sixty miles upstream; the only source of supplies to the up-river tribes.'[36]

The presence of Chinese traders in the bazaars of the various rivers set in motion the processes of economic transformation. Native collectors brought down quantities of *gutta*, *engkabang*, rattan, and other products for the Chinese shopkeepers. These products were in turn shipped to Kuching and Singapore. Thus, remote longhouses in the interior of Sarawak became enmeshed in the wider regional economy (see Figure 2) stretching all the way to Singapore, first, as collectors of jungle products and, secondly, as purchasers of ceramics, luxury goods, and foodstuffs (see Figure 1). The shopkeepers were well aware of the fact that their livelihood and prosperity depended upon the industry of the indigenous collectors of jungle produce, and it was in the traders' vested interests to keep this procurement trade going. At the same time, the native collectors relied upon the

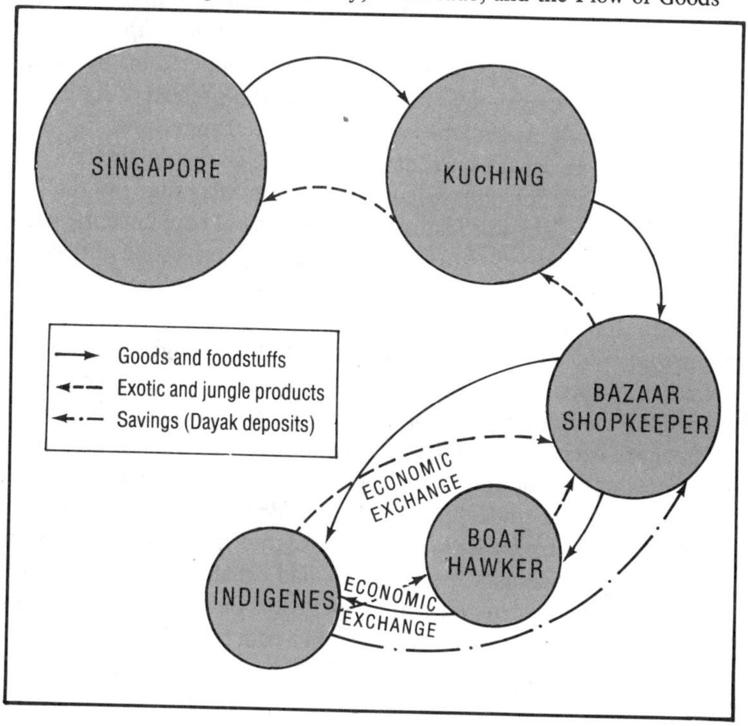

Figure 2 The Regional Economy, *Ulu* Trade, and the Flow of Goods

traders to buy their products. This relationship of mutual dependence was to characterize *ulu* trade.

The Transformation of the Riverine Economies

The relentless extraction of jungle products by the indigenous population for sale to Chinese traders transformed the riverine economies from self-contained economies based on subsistence farming into those based on exchange and inextricably linked with external markets, and with increasing sophistication in the use of money. Trading was a dynamic activity which was to drastically affect the indigenous way of life as the locals became increasingly more involved in it.

In pre-Brooke days, exchange was based upon local standards of currency. Brunei traders who were active in the coastal areas of pre-Brooke Sarawak introduced the brass cannon as currency. The cannons, manufactured in Brunei, were between 5 and 10 feet long. Each cannon was valued according to its weight (at a given amount per unit weight of metal), and then exchanged for other articles. Apart from brass cannons, other items to serve as a medium of exchange included *blachu* (black cotton cloth), *kain asap* (nankeen cotton), and small bits of iron.[37] The issue of copper coins by James Brooke and the arrival of Chinese traders, who preferred the use of money, gradually forced the old units of exchange into oblivion. James Brooke based the currency of Sarawak on the dollar standard of Singapore. He introduced copper coins in 1842 and, in 1863, a new copper currency was instituted, consisting of copper cents, half-cents, and quarter-cents. In 1900, a silver coinage consisting of twenty-cent, ten-cent, and five-cent pieces was introduced.[38]

The coastal bazaars, settlements, and longhouses were among the earliest to use copper coins as a medium of exchange. Spenser St. John, private secretary to James Brooke, commented on its increasing use among the Land Dayaks of the First Division. In 1862, he wrote: '[They] are exceedingly quick in commercial transaction, and most of them who did not know of the value of a piece of money six years ago are now active traders....'[39] Charles Brooke, whose intention it was to promote trade, had, by 1891, ordered the issue of $330,000 worth of copper coins.[40] As the coins had ornamental value for the Ibans, the Rajah instructed that copper coins be minted with a small round hole[41] in the centre of each so that a coin could be hung around the neck on a string or several coins could be joined together to form belts.

Trade based on money as a medium of exchange, as opposed to barter trade, was a principal cause of structural economic change in rural Sarawak. It affected indigenes even in remote areas. In the Batang Lupar, Ibans from as far away as its headwaters (Batang Ai) would make the long boat trips to Simanggang, 50 miles down-river. In 1890, in one such trip, the Ibans brought down $400 worth of *gutta*:

> The Ibans, for many months past, have been actively engaged in the search for *gutta* and *rattans*. During the past month, parties of them have been bringing down the fruits of their labour to Simanggang. Some boats have as much as eight pikuls. Many of these Ibans had never come down so far in their lives before, and they obtained good prices for their *gutta*, being pleased with their visit.[42]

The collection of forest products for sale to the Chinese disrupted the traditional Iban activity of cultivating rice. Some Ibans left their farms to go in search of valuable seasonal commodities such as *engkabang*. As jungle products could be collected and used to buy prestigious old earthenware and ceramics, the social concept of *bejalai* (to journey) was glorified. Adventurous young men scoured the jungles of Borneo, and even went as far afield as Malaya, Sumatra, and New Guinea, in search of valuable *gutta*, *damar*, and other sought-after jungle products.[43]

An editorial in the *Sarawak Gazette* of July 1889 commented on Iban enterprise:

> The Ibans of Saribas, the great headhunters of former times, now give their attention to hunting for jungle produce as a means whereby they may become rich enough to acquire the old jars they so highly prize. A party of 31 of this tribe left here on 1st June by the S.S. *Normandy* for Sumatra via Singapore. There they intend to look for *gutta*, and speak of Menang Kabau as the place from which to start. Another party of the same tribe will follow shortly, headed by Minggat and his son, their head chiefs, who are rich and will take funds to buy up jars. The Saribas Dayaks are firm, able bodied men, and can well afford to trust to strength of limb and endurance in any climate.[44]

As the involvement of Ibans in the procurement trade grew, there were indications that they were becoming dependent on Chinese traders, not only for old jars and artefacts, but for foodstuffs as well. Rice shortages were common by the beginning of the twentieth century in the resource-poor Batang Lupar region, mainly due to a change in emphasis from padi-growing to the gathering of jungle products. In 1904, there was a serious rice deficit in the Krian and Batang Lupar districts: 'The Chinese *bandongs* that returned early in the month from Kuching to Lingga and Simanggang brought large cargoes of rice and the price fell to 40 cents a *gantang*. There was a great rush by the Malays and the Balau Dayaks to buy at this rate.'[45]

In 1911, there was a rice shortage throughout the Second Division, and this was attributed to native involvement in the procurement trade. In 1912, it was estimated that while three-quarters of the Iban population of the Batang Lupar were busy on their farms, the rest were occupied in gathering *engkabang*.[46] In 1911, the Resident of the Second Division submitted a request to Kuching asking for a decree to prevent the export of padi from that Division; he believed that Chinese shopkeepers were buying up all the padi they could find and shipping it off to Kuching to sell.[47] The Resident was anxious to ensure that sufficient stocks of rice were available in the Simanggang bazaar for purchase by the natives involved in the procurement of jungle products.

Almost a decade later, in 1920, there was another shortage, but this time, the government took the preventive measure of buying up all surplus rice stocks in the Simanggang bazaar:

> This step has been appreciated by those dependent on the bazaar for their supplies, on which the shopkeepers were making a profit of anything up to 100%.... The native rice comes to the bazaar by trading boats, [and the shopkeepers] receive the same in small quantities from the Dayaks by barter for salt, tobacco and oil etc. The Chinese are making profits both ways. It is a great pity that the Ibans should sell their rice early in the year, but the custom is difficult to stop as they for the most part have no other mode of exchange.[48]

The dependence of the Batang Lupar Ibans upon Chinese traders was reflected in their having to sell rice, their staple diet, to the Chinese in the bazaars. This dependence came about as a result of their participation in the regional economy. But the effects of the procurement trade were unevenly felt in different river areas.

The Rejang River basin was richly endowed with forest products, and there was strong competition between Chinese traders for these. Migrant Ibans and Kayans of the Rejang River basin were active collectors of jungle produce. Boat parties from up-river would sail down to the Belaga and Kapit bazaars bringing bundles of rattan, *gutta*, and other jungle products to sell. In 1879, F. D. de Rozario, officer-in-charge of the Balleh station, reported: '... two or three boats are coming in continually from the Balleh and Balui for our bazaar. There is scarcity of cash among the Chinese to go against the

Ibans for their *gutta*. The Ibans are obliged to make an exchange of *gutta* in cloth and brassware....'⁴⁹

The Balleh bazaar did not have enough cash to purchase *gutta*, and in 1915 and 1920, the Simanggang bazaar ran out of cash with which to buy *engkabang*. Shortages of cash in the bazaar illustrated a number of points. The indigenes had not yet learned to utilize cash as a medium of exchange, and the official belief was that the natives, in general, liked to hoard copper coins.⁵⁰ For Ibans and other indigenes, the preferred mode of trade was probably barter exchange. Another reason for the shortage of cash could have been due to the Chinese traders not being backed to the hilt with available credit from Kuching or Singapore. Despite these adverse factors, the Chinese sailing junks were not discouraged from going up-river as far as Kapit. Two small Chinese junks, weighing 11 and 15 tons respectively, sailed from the Upper Rejang for Kuching in November 1898, carrying cargoes (principally *gutta*) valued at $4,420 and $10,710.⁵¹

Competition among the Chinese traders for the products brought down by the natives was so great that as soon as the natives moored their boats to the landing stages, enthusiastic traders would literally drag the bewildered Ibans and their goods to their shops. Occasionally, unpleasant accidents occurred. In 1877, two traders, Ah Kiong and Ah Bi, were each sentenced to two months' imprisonment for capsizing an Iban trading boat, causing the Ibans to lose their produce. Six others were fined $10 each for the same offence and were ordered to compensate the Ibans for the value of goods lost.⁵²

The up-river bazaars of Kapit and Belaga attracted Kayans from as far away as the eastern side of the watershed, from the Batang Kayan, Batang Mahakam, and Batang Bulungan. In June 1899, a party of 'Uma Kulit' (Kayans) arrived in Kapit on a trading visit from the Batang Kayan.⁵³ The group numbered 250 people in thirty small boats.⁵⁴ Their arrival aroused concern, and the Resident of the Third Division, H. F. Deshon, worried for their safety during their return journey. During the same month of June 1899, a Chinese trader, Chong Moiw, arrived overland from the Mahakam River, accompanied by two Malays and a party of eighteen Kayans, bringing *gutta* for sale.⁵⁵ Their arrival excited the local traders who were eager to purchase the *gutta*. The visitors subsequently complained of the theft of their *gutta* along with newly purchased goods. The theft was attributed to the Chinese who swamped the visitors in their eagerness to buy the latter's *gutta*. As it was not possible to identify those responsible for the theft, the Resident suggested that the Chinese traders compensate the Kayans for their losses, reasoning that it was in their interest to keep on good terms with the sellers.[56] The influx of Kayans from across the watershed, showing a preference for trade in the bazaars of the Rejang River basin, continued with the arrival of a trading expedition of fifty-five Kayans from the Batang Bulungan in 1906:

They brought thousands of dollars worth of good quality *gutta*, and took back large quantities of salt, piece goods and hardware. In spite of the great distance they had come, they stated that as long as they were on good terms with the *hivans* (Ibans) they found the Rejang markets accessible and preferable to those nearer their own country.[57]

Belaga, in the heart of Kayan country, enjoyed a thriving trade with tribesmen from the other side of the watershed as well. In April 1910, an overland party of some 200 Kayans arrived in Belaga from the Batang Kayan. They disposed of 100 *pikul* of Indian rubber (*gutta percha*), and took back with them lighter but valuable products such as bezoar stones and rhinoceros horns, instead of the old beads they had hoped to procure in exchange but which were unavailable.[58] The Kayans were well aware of the monetary value of exotic products like bezoar stones and rhinoceros horns, and were willing to accept them as a substitute for ancient beads, Kayan symbols of wealth and prestige.

The Upper Rejang bazaars served a clientele that extended over the watershed. These bazaars formed part of an extensive network which stretched all the way to Sibu, with boats from Belaga taking their products down to Kapit and Sibu. In 1919, Awang Mat Nor, a Native Officer, escorted a number of Chinese and Malay traders in a contingent of thirty-eight boats, with their cargo of illipenuts and rattan and other jungle produce, from Belaga to Kapit.[59]

The procurement trade in the Rejang River basin did not cause any major dislocation in the domestic lives of the indigenes, in the way it did with the Batang Lupar Ibans who had to sell their staple crop of rice. The difference was due to the abundance of natural resources in the Rejang River basin, as compared with the scarcity in the Batang Lupar basin.

A frontier region similar to the Rejang was the Baram River basin. The rapid transformation of the Baram economy would not have been possible without the active participation of the Kayans, Kenyahs, and Ibans in gathering and selling jungle products to the Chinese traders. The Baram River basin was different from the Batang Lupar and the Rejang basins: initially, as an autonomous river basin, there was active local opposition to participation in bazaar trade under the jurisdiction of the Rajah.

In 1885, a Sebop (Kayan) chief of the Upper Tinjar branch of the Baram River, Aban Jau, and his retinue visited the Marudi bazaar, and hopes were expressed that this might lead to the opening up of the Tinjar River to trade.[60] Any government hopes of exploiting the wealth of the Upper Tinjar were foiled by Aban Jau, who refused to accept the Rajah's jurisdiction over his territory. Aban Jau was described in government records as a 'renegade, flying his own flag', and refusing to let any trader into 'his country' without first paying an entrance tax as well as duty on the produce extracted. In order to mark his boundary, he was said to have strung a rattan across the river at Long Batan, where his boundary began. From the rattan were hung various signs warning travellers of the consequences that would result if any were foolish enough to violate Aban Jau's territory or laws.[61] Aban Jau threatened to attack Marudi between 1885 and 1891, although he failed to do so on both occasions. However, before he died in November 1891, he appeared reconciled to the idea of accepting Brooke's jurisdiction over his territory. Nevertheless, the Tinjar River basin was unsafe for trade for a number of years. It could boast of no Chinese bazaars, and boat traders travelled along the river at considerable risk to themselves.

With the exception of the Tinjar River, other up-river areas were active participants in riverine trade. Like Kapit and Belaga on the Rejang River, where indigenes from the east coast of the island freely crossed the watershed, Marudi similarly attracted Kayans from the Bulungan River. On 3 February 1889, a party of Kayans under the leadership of Laki Luhob paid a visit to Marudi 'with the intention of making peace with the Baram people and to trade in this river'.[62] From the Kelabit Highlands in Bario, a party of twenty-six Kelabit chiefs and their followers arrived in Marudi in 1898: '[Their presence caused] a considerable stir among the Kayans and Kenyahs of this district.... I [C. Hose] helped them to sell their *gutta* in the bazaar for the usually high price of $105 per pikul, which greatly surprised them when they found out what they could do with their money.'[63] At a peace meeting between the Tinjar and Batang Kayan groups of Kayans and Kenyahs in Marudi in 1900, a quantity of *gutta* and other jungle products, valued at between $15,000 and $20,000,[64] was brought down for exchange.

As trade in the rivers of Sarawak expanded, so too did the natives become more skilful in the intricacies of economic exchange. Barter exchange, the use of money, and the credit system were factors responsible for transforming the riverine economies.

The System of Credit

In all three riverine basins, the natives reacted to the economic forces of trade and exchange and many took advantage of these. One mechanism which came to be understood and used by the Ibans, Kayans, and Kenyahs, was credit. The significance of credit in the expansion of riverine trade, where the use and availability of money was limited, has been discussed elsewhere.

The interdependent links in the credit chain stretching all the way from Singapore, to the Kuching merchant, to the *ulu* trader, and finally to the interior longhouses, meant that the system of trade was fairly vulnerable, as one weak link in the chain could affect the rest. The Batang Lupar, as an economically depressed region, suffered a sudden demand on its resources when, in 1914, a certain Kong Aik Bank collapsed in Singapore:

It appeared that some Kuching *towkays* had been obliged to find money to bolster up their affairs and had therefore summoned their debtors in the outstations.... Traders here are dependent entirely on raw produce. As trade has not been flourishing lately, they are hard put to it to provide the necessary payments, and in any case, the sudden call on their resources will be bound to affect trade, as money is scarce in the bazaar.[65]

If Simanggang shopkeepers were put under pressure to pay their debts to their creditors in Kuching, similar pressure could be applied on the natives who took goods in advance, promising future payment in kind, in jungle products. When the traders in the *ulu* bazaars gave out little credit, or kept pressuring their debtors to pay, the overall effect was a reduction in trade and this occurred in Simanggang in 1905 and 1915.[66] In order to generate more business, the shopkeepers resorted to making use of cash savings deposited with them by Ibans successful in the procurement of jungle produce. This again was hardly a secure arrangement for the Ibans because, when trade was at a low ebb, the Ibans were unable to recover their savings. Bailey, the Resident in the Division, described the situation somewhat unsympathetically:

As it is now, an out-station shopkeeper gets goods and credit from his Kuching *towkay* probably without any security whatever, and if he finds trade going badly with him, proceeds to borrow money locally. The poor unfortunates in the district, seeing that he has a fine shop and quantity of goods, naturally consider that he has a lot of money behind him, whereas he has perhaps, merely the credit he has obtained from his *towkay*, so that when the crash comes both the *towkay* and the locals are badly swindled.[67]

The constant fall-off in the Batang Lupar riverine trade caused many traders and boat hawkers, in debt to their creditors and unable to recover what the Ibans owed them, to abscond. There are numerous examples in the records of attempts, successful or otherwise, to abscond. In 1894, in down-river Pusa in the Saribas River basin, a Chinese trader, Ah Jim, who owed money to some Malays, tried to escape to Oya but failed. All he had was a small sailing boat, and he was a total bankrupt.[68] A shopkeeper, Kuay Chow Meng of Lingga, owed $4,200 to Kuching *towkay* when he died in 1895.[69] A Simanggang boat hawker, Lim Kim Soh, fled in 1901 to Dutch Borneo with his Iban wife and child when he was unable to repay his debts.[70]

While trade in the Batang Lupar was poor, the richly endowed Rejang River basin was more fortunate. It has already been stated that the Rejang traders competed with each other in giving out advances to Ibans and Kayans. In a single year, 1892, the Chinese traders at Kapit gave out $35,000 in advances to the Ibans, leading B. A. Bampfylde, the Resident, 'to doubt if the Chinese would be able to recover half this amount'.[71]

The competition among the traders was such that up-river traders in Kapit and Song were not allowed to deal with natives who had taken advances from down-river traders at Kanowit. *Towkay* usually had regular clients with whom they conducted business, and it was rare for Ibans to switch *towkay* unless they were looking for a way out of their debts. There was always a temptation for shopkeepers to buy products from sellers other than their established clients, regardless of the fact that these natives might be indebted to other *towkay*. It was to prevent court litigation involving competing traders that the Brookes laid down an unwritten rule that up-river traders should not interfere with clients of down-river shopkeepers, as the following court case exemplifies. In this 1922 case, a shopkeeper was fined for intercepting the goods of an Iban who was already indebted to another trader:

In Chop Soon Ho Lee v. Chop Poh Soon and Nili (Dayak), the plaintiff stated that Nili came to him from Ngemah and said he was going to work in the Katibas and wanted advances from him and he gave him $116 on the condition that the *gutta* collected was to be sold to him and to no one else. After two years, the defendant returned and his *gutta* was detained at Song by

Chop Poh Soon who took 32 pieces of *gutta*. Chop Soon Ho Lee received 65 pieces which has settled the debt except for $20.... The court verdict was that Nili had no right to allow his people to take advances from a Chinese at Kanowit. The trading custom up-river with Ibans is that they can sell their produce to the Chinese who has advanced them the money or goods. The defendants were fined $10 each as a warning with costs of $1 each.[72]

Besides competing with each other in the 'interception' of their clients' product, the Chinese traders had to contend with Malay boat hawkers. In 1900, various Kapit *towkay* complained of some Malays who travelled up-river and bought *gutta* from Ibans who had already been given credit by the Kapit shopkeepers.[73] This practice went on for some time; again, in 1910, the Kapit Chinese petitioned the Resident, Baring-Gould, and asked that severe measures be imposed in cases of Ibans selling *gutta* to Malays when they were still in debt. A similar petition was presented to the Rajah when he presided over the Sibu Court. Charles Brooke's reply was that the 'Dayaks lived far away in the upper reaches of the river and if they, the Chinese, gave advances to such people they did so at their own risk.'[74] The granting of credit was crucial to the expansion of the procurement trade, but the government refused to be drawn into litigation which would entail practical difficulties in summoning the Ibans from the interior. The Rajah, in a letter to one of his officers on the coastal station of Bintulu, voiced his opposition to the idea of summoning the natives for debts:

The officer of the station has perfect right to call on Ibans owing Chinese or others money when they visit a station, to pay any just debts without any summons at all. The Punans of Pandan (35 miles away from Bintulu), I think, quite come within reach of the bazaar for summons. At the same time, it will be a mistake, in many ways, to encourage the Chinese making advances to these people—this they do often in the most heedless manner and deserve to lose their money—so in future you can issue summons to the nearby Punans.[75]

While many Rejang Ibans were indebted to Chinese traders down-river, the more successful ones in the collection trade would accumulate their savings as 'deposits' with shopkeepers. In 1902, the Resident of the Third Division registered 60 loans totalling $10,000, and Kapit traders had about $20,000 worth of deposits from Upper Rejang Ibans.[76] Hard work and enterprise had brought wealth to some natives, and those who knew how to handle their money invested their savings by purchasing bazaar shophouses, in the same way that they bought old ceramics. This was unlike the Batang Lupar Ibans who might deposit their savings with shopkeepers but who did not invest in shophouses. The shophouses owned by the Rejang Ibans were usually rented to Chinese traders. The sophistication of the Rejang Ibans in handling money was matched by that of the Baram Kayans and Kenyahs.

The system of credit in the Baram district was similar to that in the Rejang River basin and the Batang Lupar. A weak point of the system in the Baram, as in the case of the other two river basins, was that of debtors absconding. In the Baram, debtors absconded not because they were unable to pay up and were desperate, but because the Kayans were knowledgeable about trade practices and were aware that the Brooke legal structure worked to their advantage. Charles Hose, the Baram Resident, in a letter to Kuching written in 1891, described how trade based on credit worked in his district. It was a hypothetical account, but was impressively accurate in describing the way in which credit was granted:

To begin with, the Kuching *towkay* supplies his Baram agent with goods, he expecting to receive jungle produce in return for payment. The Baram shopkeeper, in return, barters the goods with those who come to buy at his shop, but also in order to obtain larger profits, and better returns, he sends the greater part of his stock in trade up-river, in boats. Small traders, mostly Malays, come and ask him for goods, sometimes to the amount of $400 agreeing to pay him in rattan or *gutta percha*, to be taken at a lower value than the current rate in the bazaar.

The Malay trader then starts up-river in his small boat with his small stock of goods, often intending honestly to pay the shopkeeper someday when he has sold the goods, and get back their value, with his own profit, for taking the goods up-river. At his first arrival at the first house he stops, Kayans come down, overturn his goods, examining everything. Some buy from him promising to pay in rattans, on his return to the house. Those who don't come to buy, beg tobacco, and he gives them tobacco because he is obliged to keep on good terms with them, not because he has any friendly feeling towards them, for he hates the sight of them, and so he pushes on up-river stopping at each house, spending from one month to three months and making arrangements to receive rattans, which are, as the Kayans tell him, already cut in the jungle and only waiting to be carried to the river. On his return journey he begins to find that most of the people who promised to go and fetch the rattans a month before have not moved out of the house all the time. With what few rattans that have been paid over to him, he is now obliged to return to the bazaar to keep his promise as far as he can with the Chinese shopkeeper, and as he has disposed of all the goods allowed by the shopkeeper, he must obtain more in the same way to earn a living. The shopkeeper sees how things are, and often allows him more credit. If however the shopkeeper will not give him more credit, he goes to someone else, in which case he then starts away up-river again; this time he is able to collect something on the old account, but by no means is fully paid up, and those who purchased of him before, ask for more credit. Whilst he is away, the shopkeeper he formerly took goods from, hears he has borrowed goods from someone else and is afraid he will now only get back a part, each time the Malay trader returns to the bazaar he keeps back a part for the shopkeeper who gave the more recent advance.

Then the man who gave the original advance to the Malay sees it will be a long time before he gets his credit back, and to ensure payment being made to him, takes out a summons in the Court, against the Malay. The case comes up for a hearing, generally very few things are disputed, and the defendant is given time to liquidate his debt. He goes up-river this time with little or, nothing to trade with, only hoping to collect his credit. Some Kayans, seeing his unfortunate position, make an effort to settle with him; others, often small chiefs or heads of houses, are rather pleased to hear he has been summoned, and make no effort whatever, hoping (if he will not give them more credit) that he may be imprisoned or that he may go elsewhere, in fact, that anything may happen to him that will delay them paying him what they know they owe him, they themselves feeling safe that a summons will not be served on them.

Now the same people who are in debt to this Malay trader, also owe a dozen other traders in the same way and often even when they have rattans send or bring them down to the bazaar to obtain goods from the shopkeeper without the slightest consideration for the up-river trader.... If however, the Malay trader is paid half his credits his profits are so large that they will cover his bad debts. This causes the Kayans to come to the bazaar to buy and, finding things so much cheaper, consider themselves swindled by the up-river trader.[77]

Hose was referring to Malay traders, possibly Brunei Malays who were gradually ousted from the procurement trade by the Chinese. The boat traders in question could very well have been Chinese, many of whom took up boat hawking. From this account, it can be seen that Kayans took advantage of the fact that they could not be summoned for debts. Yet the boat hawkers could do little except give out more credit in the hope that they might be able to recover a percentage of it. If the Kayans refused or were unable to pay, it meant financial disaster for the boat trader. The Kayans also preferred to take their rattan and *gutta* to the bazaar, even if they were already indebted to up-river traders. They may have done this for a number of reasons: as a form of negative sanction on the up-river traders who, they believed (rightly or wrongly), had cheated them and, secondly, to benefit from a legal system which purported to protect their interests. The Brookes had clearly under-estimated the Kayans' ability to understand the intricacies of trade.

As early as the time of the founding of Marudi in 1882, it was said that 'considerable annoyance and loss was caused to the Claudetown (Marudi) traders by their debtors absconding to Brunei territory'.[78] A perusal of the traders' books in 1894 revealed that debts owed to the bazaar shopkeepers by absconders to Belait amounted to $1,629.68.[79] Some traders, not unlike their clients, resorted to absconding. They behaved in the same manner as bankrupt traders in the Batang Lupar district. In 1904, Lai Ghee, a Baram shopkeeper owing about $2,000 in the Debtors' Court to Chinese firms in Kuching, tried to escape to Belait. However, he was caught and taken before the court, where he explained that 'he had been thinking about his debts, and felt he could never pay them ... and came to the conclusion that the best thing to do was to run away'.[80] He had tried to collect his debts from up-river Kayans but they were unable to pay him.

The procurement trade brought about economic exchange, first based on barter, then replaced by the use of money and credit. It was trade which transformed the autonomous, self-contained, subsistence economies of the riverine basins, causing the natives to depend upon Chinese traders for prestige items, luxury goods, and foodstuffs.

1. Charles Hose, *Natural Man: A Record from Borneo* (London: Macmillan, 1926; reprinted Singapore: Oxford University Press, 1988), p. 120.
2. Ibid., p. 113.
3. I. H. Burkill, *A Dictionary of the Economic Products of the Malay Peninsula*, Vol. 2 (London: Crown Agents for Colonies, 1935), p. 1875.
4. *SG*, 27 September 1877, p. 71.
5. *SG*, 1 March 1889, p. 40.
6. *SG*, 1 February 1894, p. 27.
7. *SG*, 2 April 1912, p. 77.
8. *SG*, 14 April 1914, p. 96.
9. *SG*, 1 May 1915, p. 105.
10. *SG*, 16 May 1919, p. 110.
11. According to the oral traditions of the Cantonese in Sarikei who were pioneer settlers of the Rejang River basin in the 1880s, timber was cut and exported to Hong Kong. This has been confirmed by documentary evidence in the *Sarawak Gazette* (see *SG*, 24 January 1871, p. 38; and 2 December 1889, p. 164). Oral information was obtained from Chieu Pak Ming, 20 June 1981, Sarikei.
12. *SG*, 1 September 1917, p. 224.
13. See Harry De Windt, *On the Equator* (London: Cassel, Peter and Galpin, 1882), pp. 58–9; and *SG*, 24 January 1871, p. 38.

14. *SG*, 1 February 1886, p. 24.
15. *SG*, 1 June 1886, p. 93.
16. *SG*, 2 August 1886, p. 123.
17. *SG*, 1 May 1891, pp. 68–9.
18. *SG*, 15 March 1873, p. 24; and 24 November 1874, p. 3.
19. Charles Hose and William McDougall, *The Pagan Tribes of Borneo*, Vol. 1 (London: Macmillan, 1912), p. 153.
20. *SG*, 16 July 1909, p. 157.
21. *SG*, 1 February 1881, p. 102.
22. *SG*, 16 April 1919, p. 91.
23. *SG*, 1 July 1884, p. 76.
24. Ibid.
25. *SG*, 31 March 1886, p. 56.
26. *SG*, 1 March 1887, p. 45.
27. Ibid.
28. *SG*, 12 May 1887, p. 90.
29. *SG*, 1 April 1897, p. 74.
30. *SG*, 2 August 1897, p. 148.
31. *SG*, 2 March 1905, p. 57.
32. *SG*, 16 July 1909, p. 158.
33. *SG*, 16 May 1910, p. 109.
34. Ibid.
35. *SG*, 2 April 1912, p. 74.
36. Charles Hose, *The Field-Book of a Jungle Wallah* (London: H. F. & G. Witherby, 1929; reprinted Singapore: Oxford University Press, 1985), p. 21. Hose was the Resident in Marudi.
37. See Frederick Boyle, *Adventures among the Dayaks of Borneo* (London: Hurst and Blackett, 1865), p. 100; and W. H. Treacher, 'British Borneo: Sketches of Brunei, Sarawak, Labuan and North Borneo', *JSBRAS*, No. 21 (1890), p. 43.
38. I. Pridmore, 'Sarawak Currency', *SMJ*, Vol. 7, No. 7 (1956), p. 112.
39. Cited by Henry Ling Roth, *The Natives of Sarawak and British North Borneo*, Vol. 1 (London: Truslove and Hanson, 1896), p. 231.
40. *SG*, 2 February 1891, p. 17.
41. Letters of Rajah Charles Brooke, Vol. 2, 14 March 1891, p. 178, SA.
42. *SG*, 1 August 1890, p. 100; and 1 September 1890, p. 118.
43. See Chapter 4, note 95.
44. *SG*, 1 July 1889, p. 95.
45. *SG*, 3 March 1904, p. 54.
46. *SG*, 2 April 1912, p. 77.
47. The Rajah's Confidential Letter Book, 1911, Vol. 1, 1 October 1911, SA.
48. *SG*, 16 July 1920, p. 163.
49. F. D. de Rozario to A. R. Houghton, Resident of Rejang, Letters, 28 February 1879, SA.
50. *SG*, 16 May 1919, p. 118.
51. *SG*, 1 November 1898, p. 208.
52. 'Government v. Lai Choon, Ah Bi, Kim Seng, Ah On, Ah Leng, Ah Hing, Ah Teh and Chunnio', 3 February 1877, Sibu CCB, Vol. 3, p. 146, SA.
53. *SG*, 1 June 1899, p. 121.
54. *SG*, 1 July 1899, p. 240.
55. *SG*, 1 June 1899, p. 122.

ORGANIZATION OF ULU TRADE 121

56. SG, 1 August 1899, p. 226.
57. SG, 16 February 1906, pp. 59–60.
58. SG, 1 June 1910, p. 118.
59. SG, 2 June 1919, pp. 146–7.
60. SG, 2 March 1885, p. 24.
61. Hose, Natural Man, p. 49; and SG, 31 March 1968, p. 57.
62. SG, 1 May 1889, p. 72.
63. SG, 1 June 1898, pp. 121–2.
64. SG, 1 May 1900, p. 99.
65. SG, 16 April 1914, p. 96.
66. SG, 3 March 1906, p. 56; and 1 April 1915, p. 77.
67. SG, 2 November 1905, pp. 255–6.
68. SG, 1 March 1894, p. 44.
69. SG, 1 February 1895, p. 38.
70. SG, 2 January 1901, p. 12.
71. SG, 1 December 1892, p. 218.
72. 'Chop Soon Ho Lee v. Chop Poh Soon and Nili', 7 November 1922, Sibu CCB, p. 99, SA.
73. SG, 2 April 1900, p. 75.
74. SG, 16 September 1910, p. 199.
75. C. Brooke to D. Owen, Letters, Vol. 9, 10 February 1903, pp. 102–3, SA.
76. SG, 3 March 1903; and 2 May 1903; as cited in Robert Pringle, *Rajahs and Rebels* (London: Macmillan, 1970), p. 294.
77. SG, 1 February 1892, p. 35.
78. SG, 1 June 1882, p. 78.
79. SG, 1 November 1894, p. 191.
80. 'Government v. Lai Ghee', 25 May 1904, Baram CCB, Vol. IV, p. 67, SA.

6
Up-river Trading and Social Relations

WHERE and how the *ulu* traders conducted their trading activities with the indigenous groups was a subject of much concern to the White Rajahs and their officers. The Brookes, while giving encouragement to the Chinese in their role as traders, were anxious to regulate and control their movements and activities.

Trading Regulations and Restrictions on the Chinese

There were a number of reasons for Brooke anxiety over the *ulu* Chinese traders: the authority of the Rajah was still to be established in some up-river areas; control over trade also constituted an effective sanction against 'rebellious' up-river Ibans, Kayans, and Kenyahs; finally, in keeping with administrative considerations, the Brookes preferred the Chinese to stay in bazaars.

There was no conscious overall policy towards Chinese traders, but Residents and their officers had clear instructions on how to deal with them on certain matters. The Residents and their subordinates were particularly concerned with Chinese living and trading in Iban longhouses, with the movements of boat hawkers, with the enforcement of trading regulations, with the way traders built and maintained their shophouses, and with the collection of shophouse assessments and other taxes and licences.

One of the most important Brooke 'laws' which has had a lasting impact on social relations was the unwritten rule that no Chinese or Malay could live and trade in native longhouses. The Brookes, often fighting for their own survival, were concerned that the different races be kept separate, and as trade entailed much social contact, it was pertinent that it be confined to the bazaar or boat. There was also a paternalistic desire to prevent the Ibans from being swindled. Moreover, administrative convenience was in the minds of the Brookes. If the Chinese were restricted to the bazaars, locating them for court cases would be easier than if they were scattered along a river. Finally, it would be difficult to keep a check on the activities of the traders if they were allowed to reside freely in longhouses well beyond the reach of Brooke authority. The Brooke officers were particularly concerned about the boat hawkers with whom they had little direct dealing.

It became a finable offence for traders to 'live among Dayaks' in the late 1870s,[1] a law rigidly applied in the Batang Lupar and Rejang River for detected offenders. The normal penalty for offenders who were caught in the Batang Lupar was six months' imprisonment. In one such case, 'Government v. Chau Soon', 5 July 1882, the defendant was reported to have been living in Sungei Pinggai in the Undup branch of the Batang Lupar. During the visit of Deshon, the Resident, to the Undup River, Bulih, the headman of the longhouse at Sungei Pinggai, complained about the presence of Chau Soon:

Ah Soon denies that he has been living at Sungei Pinggai and produces a *surat mudai* [document] dated 7th May 1880. From this it would appear that he has ever since been in the Undup under pretence of trading as he has not returned the pass and is therefore liable for two years and two months' duty at the rate of 18 cents per month.... The defendant is sentenced to six months' imprisonment for living in a Dayak house in defiance of government orders.[2]

The mouth of the Undup River is only 5 miles above Simanggang, and yet this particular trader had been living and trading in an Iban longhouse without the knowledge of Brooke officials. For over two years, he had not renewed his pass, nor was he asked to until accidentally caught by the Resident. As traders depended on the hospitality of their hosts in whose longhouses they stayed, they had to be on their best conduct or risk being informed upon by their clients.

In the Rejang River basin in 1878, 'Ah Toon was fined $20 for neglecting to comply with the order of the government directing all Chinese living in Dayak longhouses to return to (Sibu) bazaar.'[3] The headman of the longhouse had requested that Ah Toon be got rid of as he could no longer vouch for Ah Toon's safety. The court case stated: 'Ah Toon is represented to be a quarrelsome neighbour, scurrilous and also unfair in his dealings.'[4] Fraudulent trading was the reason for the Ibans' request for Ah Toon's eviction. The Ibans, on their part, were continuously warned against letting Chinese traders stay in their longhouses. An Iban headman in the Batang Lupar who gave permission to a trader to live in his longhouse was fined 15 *kati* of padi in 1914.[5] Another headman was dismissed in 1922 for keeping four Chinese traders in his longhouse after dark.[6]

There was a difference in treatment of offenders between the Lupar and Rejang River basins. Offenders in the Batang Lupar were imprisoned, while those in the Rejang were fined although, by the

1920s, jail sentences were also being meted out in the Rejang.[7] The geographical expanse of the Rejang River basin and limited manpower resources could have been responsible for the lenient treatment given to offenders. As it was considered a serious offence, no excuses were accepted by Brooke officers from Chinese traders attempting to set up shop in longhouses.

There were a number of reasons for these attempts by Chinese to live among the natives. If they were permanently settled, the traders did not have to row their boats from longhouse to longhouse, saving themselves considerable time and giving them an edge over bazaar and other boat traders in a competitive business. There were others sincere in their desire to become 'native', especially when they had native spouses.[8] But the Brookes were unprepared to allow any concessions. A Chinese trader who lived together with down-river Ibans in Tinggal near Lingga in the Second Division was asked to leave, and was also fined:

> The defendant has been living at Tinggal for some years in defiance of government regulations, when ordered by Sipi [under instructions from the Resident], he neglected to come down. In defence, he has only to say that he has lived among Dayaks for nearly all his life. He is therefore fined $15 and ordered to remove to Lingga or some other place among his own countrymen.[9]

These examples represent only the occasional cases of 'living among Dayaks' that were detected. In fact, the *ulu* trader was very much on his own and, though many may not have stayed permanently to trade and live in longhouses, the regulation of allowing a trader to spend a stipulated period only at a longhouse was almost certainly ignored.[10] The maximum number of days theoretically allowed at a longhouse landing point varied from one river basin to the next. While a two-day stay was the permissible limit in the Baram, fourteen days were allowed in the Batang Lupar.[11] Sometimes, the traders had no choice but to ignore the edict and stay longer if the level of the river suddenly rose or fell, thus preventing travelling,[12] or if they were waiting for the Ibans to procure the jungle products for which advances had been given.

The Brookes believed that traders who lived in longhouses would cheat the natives—which at times happened. However, the main reason was to ensure that the different ethnic groups remained segregated. Policies to this effect were enforced in all territories annexed by the White Rajahs.

In the Baram River basin, annexed by Charles Brooke in 1882, 'living in Dayak houses' did not appear as an offence as such in the court records. Rather, the offence was classified under 'going into prohibited territory without passes'. On 21 September 1887, C. W. Daubeny, officer-in-charge of the Baram, prosecuted one Moh Sin, a Teochiu trader, who was 'accused of proceeding up the Bakong River ... Moh was sentenced to one year's imprisonment with hard labour'.[13] A perusal of the court records reveals the autonomy of Baram boat hawkers who defied government orders by going up as far as the Bario highlands[14] and beyond Long Akah.[15]

Pressure was applied on the Baram up-river traders to build bazaars and settle down permanently, as was the case in other parts of Sarawak. In 1919, the Chinese traders at Long Akah were ordered 'to complete the building of a new bazaar in four months' time, and whoever has not, will be declared a bankrupt and taken down-river'.[16] Traders could go up-river as far as Lio Matu by taking out hawkers' licences, but were only allowed a maximum of two nights' stay at each *pangkalan*.[17] Regulations like this were ignored with impunity if there was no one to enforce them. An example from 1920 illustrates this point all too well: a number of traders in Sungei Tinjar were ordered in 1918 to build a bazaar, but, two years later, there was still no bazaar. In 1920, the traders were fined, and warned to finish their bazaar within two months or face the risk of being banned from trading up-river in future.[18]

According to Brooke law, boat hawkers were meant to operate only from their boats and not carry their wares up to the longhouses. While some observed this rule, there were others who either were not aware of the regulation, or deliberately chose to ignore it.[19] Traders were not supposed to stay for extended periods at the *pangkalan* of longhouses and had to be constantly on the move.

A regulation of 1900 required hawkers to come down to a government station once a month to renew their licences.[20] An example of a Simanggang boat trader who was prosecuted for not making his monthly trip down-river to renew his licence is found in 'Government v. Ah Kee' in August 1921: 'The defendant stated that he does not know that trading boats have to come down every month; he was fined $25 and was told that he must bring his *bandong* down to the river mouth every month.'[21]

The pedlar in this example could have been unaware of Brooke edicts. According to local informants, Brooke officials were a distant authority with whom the boat traders seldom had contact. The authority figures with whom they had to contend regularly were the local headmen of the longhouses: it was from them that actual permission was obtained to trade or even to stay in the longhouses. An oral source, Syn Chin Joo of Kapit, said traders were aware that goods were not to be taken into longhouses. Nevertheless, the longhouse *tuai rumah* would be pleased if the hawkers took their wares up as it was easier to trade, and 'they would be safe from plunderers and crocodiles'.[22] A Kanowit trader, Siaw Kee Siong, claimed that he usually stayed overnight or longer in Iban longhouses when he was travelling.[23]

Boat traders in the riverine basins of rural Sarawak enjoyed considerable autonomy. This was especially so in remote up-river areas where the Rajah's claim to jurisdiction was invariably in doubt. For example, the unsettled conditions of the Upper Rejang, where the Kayans and Kenyahs were by no means prepared to accept the authority of the Brookes,[24] may not have been the safest for trade, but left the traders free of 'official' interference. When government authority ultimately reached remote Belaga in 1909:

Some fifteen Chinese and Malay traders were convicted and incurred punishment for living among the Kayan, Kenyah and Punan people contrary to government orders, these traders have been repeatedly warned; they have had their huts destroyed on more than one occasion but have persisted in disregarding the government order.[25]

As late as 1939, the Belaga station continued to be visited not more than once a year by Brooke officers.[26]

While Brooke officials in the course of their occasional visits upriver were able to catch traders contravening Brooke edicts, the most common type of encounter between traders and the Rajah's men was during the traders' down-river trips for the purposes of unloading, victualling, or renewing licences. Invariably, the traders had to pass the down-river fort, from which it was possible to observe boat movements along the river (the fort was always built in a commanding position). In Simanggang, in 1883, a trader ignored the fort sentry's call to stop for an inspection and was duly warned of a future fine if he again paid no attention to orders to halt.[27]

Other restrictions faced by up-river traders concerned the selling of salt, guns, and ammunition. Restricting the sale of salt was one of the more effective ways of applying indirect pressure on up-river Ibans and Kayans. Charles Brooke, in a memo to the Resident of the Rejang in 1880, concerning security measures at Fort Sylvia in Kapit in the expectation of Kayan attacks, cautioned, 'Some watch had better be kept on the sale of salt to the Kayans in order that they may not stock any large amount.'[28] The sale of guns and ammunition was not allowed because the Brookes feared that such weapons could be used against them.

The Simanggang and Sibu court records provide many examples of traders who were penalized for excesses in salt trading.[29] The sale of guns was considered a more serious offence. In 1894, a Betong Chinese goldsmith, who had made a gun out of an old iron barrel for an Iban, was fined $2 and warned that 'any Chinaman who makes any gun for Dayaks or others in this place without permission will be liable to a term of imprisonment ... '.[30] In 1923, a fine of $50 was imposed on a Kanowit trader who sold guns to three Ibans.[31]

While the Brookes were concerned over up-river traders 'living amongst Dayaks' and peddling in salt and guns, the down-river traders in the bazaars were subjected to other kinds of restrictions and regulations. Brooke officials, seeing themselves as the custodians of native interests, were concerned over bazaar shopkeepers using fraudulent weights and measures in their transactions with the Ibans and other native groups.

Their concern was not without justification for deception in trade was widespread, judging from *Sarawak Gazette* reports, court records, and the accounts of informants. A common trick was that of using two different weights for the *datching*, one for buying, and the other for selling, with a considerable difference between them. The measuring rods for the *datching* were also sometimes short of the standard requirements.[32] Another trick was to have two *passu* containers. Yet another was for *gantang* measures to be fitted with false or concealed bottoms. A hidden bottom which reduced a *gantang* measure's holding capacity was used for selling,[33] while for buying, an enlarged bottom could be used to increase its volume,[34] without customers being aware of it. There were other more subtle ways of cheating. By far, the favourite trick was to weigh goods at night, in dim light. Another trick was to place hidden sticks protruding out of the shop-floor to support weighed produce, thus making the scales weigh lighter.[35] According to informants, the traders practised fraud to take advantage of natives who had little knowledge of trade, in retaliation against clients who did not pay their debts, or as 'compensation' for the hospitality in food and accommodation extended by traders to their customers.

Bazaars from the Batang Lupar to the Rejang and the Baram were periodically subjected to wholesale inspection of their weights and measures. In 1892, all the Saratok shopkeepers were fined for possessing false measures,[36] and in a raid on Simanggang bazaar in 1905, 'it was found on examination that every *datching* in use was false'.[37]

A search in the Sibu bazaar in 1912 resulted in the discovery of tin measures called 'company *gantangs*' of varying sizes, 'some having double the capacity of a regulation *gantang*'.[38] Fifty traders were prosecuted in court. When Song bazaar was raided in 1924, twenty 1-gallon measures were confiscated and two traders were heavily fined $135 and $125 respectively as a salutary warning.[39] Over in Marudi, in September 1896, fines totalling $49 were collected for various offences[40] involving weights and measures.

Occasionally, the Ibans saw through the trickery of the traders, as illustrated by this 1914 example from Simanggang. A major Simanggang firm, Chop Chiap Heng, was given a warning as a result of a complaint by some Ulu Ai Ibans who wanted to sell *gutta* to the shop. This was how the deception was carried out, according to the report filed by the Resident in Simanggang:

> They [the Ibans] distinctly saw the coolie who was weighing the *gutta*, make it lighter by pressure with his foot, thus making the *gutta's* weight 16 katis, whereas previously it had been weighed and found to be 20 katis.... Chop Chiap Heng is heavily censured for this. If any of this sort of thing is heard of, Chop Chiap Heng will be prosecuted. As it is, the transaction did not take place, as the Dayaks had seen through the trick.[41]

Indeed, the Chinese were not the only ones who indulged in fraudulent trade practices. As already seen in the last two chapters, native retaliation to being cheated by the Chinese included robbery, murder, and absconding from debts. However, when it came to the art of fraud, the natives were more than a match for the traders. The Ibans and other natives would either mix pieces of bark, wood, or rubbish with the *gutta* to increase its weight, or else coat it with sawdust. In Simanggang in 1895, five Ibans were fined $25 each for adulterating the *gutta* they were going to sell.[42] In another incident in 1906, five Ibans who sold mixed *gutta* to a Chinese trader, Ah Bah, of Lubok Antu bazaar, refused to answer a government summons to appear in court and took off into the jungle.[43] In Kanowit, the mixing of *gutta* was done in a more sophisticated way with the Ibans claiming the adulterated *gutta* as a new product, '*gutta baru*', which was just some *jelutong* mixed with a large quantity of red sawdust.[44] As the *gutta* trade was profitable in the Rejang River basin, some shopkeepers preferred to close an eye to the mixed product and would pass it on to the markets outside.[45] A comparison of prices paid for mixed and pure *gutta* in the Baram in 1901 reveal that mixed *gutta* fetched from $50 to $90 a *pikul*, while well-cleaned *gutta* brought in from $200 to $350 per *pikul*.[46]

Shophouses were occasionally inspected to make sure they were in habitable condition and for collection of assessment on the premises. New shops had to conform to the requirements of having kitchens separate from the main buildings so as to prevent the outbreak of large-scale bazaar fires.[47] In 1895, an inspection of the Simanggang bazaar was carried out: out of a total of 45 shops, 11 were ruined and unoccupied, while 14 others required repairs.[48] Pressure was exerted on the shopkeepers to renovate their premises and, three years later, the situation had improved somewhat, with only five shophouses remaining decrepit, and four needing repairs.[49] The Marudi bazaar in the Baram River basin was inspected in 1916 to make sure that the shophouses were sanitary. D. C. Newington, Resident of the Baram, commented:

By-laws for the cleanliness of the bazaar have been issued and the result is very satisfactory. The trouble here-to-fore has been the lack of space for storing and drying *rattans*, and in consequence of this, the Chinese have been stacking *rattans* on the five-foot way and drying them on top of ditches, thereby fouling the ditches, and also damaging the cement work by the heavy weight.[50]

In 1920, a number of traders were taken to task in the Baram court for breaching various bazaar regulations. The charges read as follows:

Choon Wai has not finished the wall of his *loteng* [top floor]; Hap Ho has no kitchen; Ah Loo has no kitchen; Swee Jak has a dirty shop; Kua Chui has no earth below his kitchen, and Ah Wang has no *loteng*.
Each of them was to pay costs of a dollar each. Choon Wai was to complete his *loteng* in ten days, Hap Ho to finish his kitchen in one month, and Swee Jak, to clean up his shop in three days. Kua Chui was to fill the earth under his kitchen in ten days, and finally, Ah Wan [sic] was to build his *loteng* within a month.[51]

A strong reason for the periodic inspections of shophouses was the collection of taxes on the premises. In 1900, a token assessment of 30 cents per month was collected from each shophouse in the Simanggang bazaar.[52] Kabong, Saratok, and Roban shophouses paid an assessment of 25 cents each. In some instances, distant bazaars paid no assessment, a reflection of their autonomy.[53] In the Rejang River basin, traders were made to pay for the security of up-river trade. Charles Brooke sent these instructions to Charles Vyner Brooke, the Resident in the Rejang, the Rajah Muda, his future heir:

I direct that the greatest care shall be taken in supervising the Ulu Ai Ibans to prevent their killing either Kayans or others:

That a boat shall be always ready to keep watch, below and above Kapit, with a crew of twenty-five Malays from the lower country commanded by Abang Ali and then Abang Mating. Abang Ali or Abang Mating will get a food allowance of $5 while engaged, and the crew which will be ordered out for three months will be allowed $3 on monthly wages and $3 paid in food—or $6 in all for each man per month.

Furthermore, I direct that the Chinese shopkeepers at Kapit, Song, Kanowit and Sibu all shall pay $1.50 per door and 50 cents for the poorer class a month for this extra armament. The first call is to be made on July 31st—and they can well afford to do this as it is for their own benefit.[54]

The regulations discussed above—that is, those concerning the prohibition on living and trading in longhouses and those regulating the number of days allowed at each longhouse landing, checks on fraudulent weights and measures, and on clean and tidy shophouses, and assessment rates—were reinforced by Brooke ethnic policies.

Ethnic Relations and Riverine Trade

The influx of Chinese settlers into Sarawak in the late nineteenth and early twentieth centuries and their taking up trade as an occupation, together with the siting of bazaars along the rivers in areas where Malays, Ibans, Kayans, and Kenyahs were predominant, not only caused economic and demographic change, but also fostered new communal relations between different ethnic groups and within the ethnic Chinese group (for example, the pioneering Teochius and Hokkiens competed with each other in business). The trading methods of the Chinese traders, their sophisticated use of money, their kinship, credit, and communication links stretching all the way to Kuching and Singapore, and the traders' constant encouragement to the natives to buy, sell, and exchange, proved far superior to the skills of the Malays who had hitherto been the traditional traders in pre-Brooke Sarawak.

In Kuching, speech-group rivalry was a dominant characteristic of Chinese society, a rivalry growing out of competition in trade between the Hokkiens, with their Chao-ann allies, and the Teochius.[55] Similar speech-group differences were pervasive in the outstations for the same reasons. The Hokkien and Teochiu dialect groups constituted separate communities. They kept to themselves, sponsoring, helping and employing members of their own speech group, and generally looking after their own interests. In matters of social welfare, such as looking after the destitute and the infirm, help did not transcend speech-group boundaries. There is a good example, dating from

1908, of the indifferent reactions of Hokkien traders in Saratok towards the illness of a Liuchiu man:

> On 6 July, Ah Chai, a Liuchiu, a fisherman from Sibu, who was working at Alit, fell violently ill of diarrhoea and was brought over by some Malays. When he vomitted, I (Suboh) sent P.C. Ambang down to the boat to ascertain how ill he was, but the policeman found him in a state of collapse. The Chinese traders took no notice of him, and after a day and night, Ah Chai died. The fortmen had to bury him, no Chinese coming to assist them as the deceased was of a different clan.[56]

Speech-group differences were more strongly asserted in open group fights in each of the three riverine basins. In 1888, six years after the permanent establishment of Marudi bazaar, there was a big fight involving about forty Hokkiens and Teochius armed with clubs. The dialect group quarrel originated in the gambling debt owed by one Ah Piow, a Hokkien, to Lai Ngee, a Teochiu:

> Ah Piow said he owed the balance of a gambling debt to Lai Ngee which he had arranged to pay, and on his going to Lai Ngee's shop to buy something, Lai Ngee demanded the money and then said to his coolies, 'Let us take it from him'. This, they tried to do, and when he struck out, Lai Ngee said he truly demanded his money and Ah Piow said he would not pay.... Ah Piow then struck Lai Ngee and one of Lai Ngee's coolies came to help Lai Ngee and the row became general.
>
> ... After some consideration, the Officer-in-charge fined the two Teochiu shops of Lai Ngee and Eng Sim $75 between them. The two Hokkien shops of Lu Wat and Yap Hin were also fined $75 between them. Ah Piow as the principal cause was fined $50, and Ah Yong of Baram, and Ah Yong of Miri, as being the first to start the general fighting were fined $25 and $10 respectively.[57]

The fight was considered serious enough to necessitate the patrolling of the bazaar by Iban rangers to prevent any more such occurrences and to allow strong dialect group feelings to cool down. That four shops, two owned by Teochius and two by Hokkiens, were implicated in the fight suggests that the catalyst of the dispute, a gambling debt, was symptomatic of general trading rivalry between the two groups bitterly opposed to each other. This pattern of speech group quarrels was carried over up-river in Long Lama in 1907,[58] and Long Akah in 1912.[59] That these mutually hostile feelings continued unabated up-river could have been because the up-river traders had as their patrons and credit suppliers the down-river *towkay*.

In the Rejang River basin, in the small coastal bazaar of Daro, serious disturbances between Hokkiens and Teochius took place in 1913[60] and 1915.[61] These prompted the Resident to take strong security measures. From what transpired in the Lower Rejang, it can be expected that the pattern of rivalry was duplicated in bazaars further up the river, where competition for trade was even greater. The Batang Lupar basin was not exempt from violent fights between Hokkiens and Teochius: in Betong in 1922, a number of Teochius set upon a Hokkien for no apparent reason.[62] Simanggang had a nearby Hakka mining colony in Marup, near Engkilili, and the bazaar Teochius strongly resented their up-river rivals. A serious affray between Teochius and Hakkas flared up in 1914 in Simanggang.[63] Like the 1888 Hokkien–Teochiu fight in Marudi, in which shopkeepers were directly or indirectly involved, some Simanggang *towkay* had to be cautioned: their acting as sureties for some of the fighters indicated their vested interests in speech-group competition.

In Kuching too, as in the outstation bazaars, street fights between Hokkien and Teochiu coolies of rival firms broke out regularly.[64] There is no evidence to suggest that these fights, either in Kuching or the *ulu* stations, were instigated by secret societies. Hokkiens and Teochius were not known to form secret organizations, though what prevented them from doing so is not clear. Perhaps their monopoly of urban and rural trade gave them little cause for dissatisfaction, except towards their competitors. Grievances among the Hokkiens and Teochius seemed to be directed towards each other, instead of against the Brookes. Rigid government supervision of the Chinese, following the 1857 *kongsi*–Brooke war, was another deterrent to the formation of secret societies, as was the death penalty for leaders of secret societies. A ready pool of Iban warriors who served the White Rajahs voluntarily also helped to ensure that Brooke rule would meet with little opposition.

Intra-Chinese fighting aside, competition in trade between Chinese and Malay traders led to the eventual demise of Malay trade and resulted in several communal clashes in the *ulu* bazaars.

By the end of the nineteenth century, there was a Chinese bazaar in almost every Brooke station, while many traders operated out of the reach of Brooke authority. The Hokkien and Teochiu traders were part of an extensive network that stretched all the way to Kuching. With their advantages of capital and organization and backing from Kuching *towkay*, the rural shopkeepers were in a position to put Malay traders out of business or, at the very least, relegate them to a subsidiary role. Malay traders operated alone or in very limited partnership and did not have the back-up or patronage which their *ulu* Chinese counterparts enjoyed. In addition, the Chinese understood the usefulness of goodwill in business and the necessity of placating their Iban and other native clients, and were willing to take risks by giving out advances of goods. Another advantage enjoyed by the Chinese was the availability of assistance from their fellow dialect speakers, kin, and families.

Although the Chinese traders provided stiff competition, the Malays were able to hold out in a number of outstation bazaars, such as Pusa and Debak in down-river Saribas, and in the up-river frontier bazaars of Kapit and Belaga. Trade rivalry was a strong factor in communal violence in many bazaar incidents. In 1885, three years after the founding of Marudi, there was a fight between Brunei Malay traders and the Chinese. The Chinese closed their premises for eight hours in protest against the incident.[65] The relationship between the Chinese and the Brunei Malays became so strained that, in 1909, the Chinese traders petitioned the Resident for the removal of Kebar, the Malay court writer, on the grounds of showing favouritism towards Malay traders and of neglecting his duties. The Malays were subsequently warned by the Resident 'not to interfere with the trade of the Chinese'.[66]

Incidents of individual assault which then became sufficiently widespread to involve bazaar and kampong occurred in Kabong in December 1903,[67] in Sibu in 1911,[68] and in Saratok in 1917.[69] The underlying cause of group fighting was communal trade rivalry, and tension ran so high that fights could be easily sparked off by minor incidents.

Economic rivalry expressed itself in frontier violence in Debak in 1928. Malay trading strength in Debak was illustrated by the fact that, while the Chinese owned only one coastal vessel running between Debak and Kuching, the Malays owned four. The Chinese traders themselves admitted they were unable to compete with the Malays as the Malays 'go to the longhouse and buy *gutta* on the spot'.[70] A major confrontation in 1928 involved almost all the bazaar Chinese and Malays in Debak:

The whole *kampong* and bazaar were concerned, but luckily no more dangerous weapons were used than sticks. Eight people were injured but no one really seriously. The Malays were chiefly to blame. Eight men were put on remand, and pledges imposed throughout the *kampong* and bazaar. Later, in the settlement of the case, two Chinese and five Malays received terms of imprisonment, and four people were fined. The Native Officer, Sidek, and the *Tua Kampong*, were dismissed by the Resident who was present during the hearing of the case, the former being proved to be largely responsible for the whole affair.[71]

On at least one previous occasion, the Chinese traders in Debak had held fears for their well-being: in 1899, after a quarrel with some Malays, they requested the Resident to allow 'no one to come to their bazaar after 8 p.m. until sunrise'.[72]

These examples of ethnic violence in the Baram, Rejang and Lupar River basins, occurred in down-river bazaars where Malays were active competitors with the Chinese, and where trading rivalry was a cause of insidious tension. However, trading rivalry did not always break out into violent fights and, in up-river Kapit and Belaga, Chinese and Malay shops coexisted peacefully.[73] The number of Malay traders up-river—as many as one-third of the shops in Belaga, for example, were owned by Malays in 1893—probably meant that they were still a potent trading community and that their share of the procurement trade was not threatened by the Chinese, hence the ability of the two communal groups to live peacefully with one another.

While the relationship between the Chinese and the Malays was marred by tension, that between Chinese and the Ibans was more accommodating. It has already been seen that, for the sake of their survival and livelihood, rural Chinese traders had to be prepared to accommodate their Iban and other native clients. To keep on good terms with the natives was in the best interests of the traders, and hospitality was extended to the natives whenever they visited the bazaars. There were kitchens at the back of the shophouses especially for the use of the native people. They could stay overnight and sleep in the lofts of shophouses when they came down to the bazaars. According to Kapit-born Syn Chin Joo, if the Ibans brought in valuable products to sell, they might be given canned sardines to eat; if not, they were given salted fish instead.[74]

There was another more positive element to this relationship between Chinese trader and native client—a willingness on the part of the trader to adapt himself remarkably to native ways. It was taken for granted that most Chinese traders could speak the Iban, Malay, Kayan, and other local languages—and many of them still do in the outstation bazaars today. Here is an example, taken from the First Division, of a trader who was thoroughly familiar and comfortable with native culture. An anonymous traveller on a visit to the Samarahan River in 1885 wrote:

I was astonished to find a Chinaman living at Santah, and by no means as a Chinaman, but as a native. To all outward appearances he was a Dayak, having adopted their primitive costume, and hiding his pigtail in an ample kerchief. All this, as the observer could make out, was done for the purpose of increasing his popularity and faculties of trade among the Dayaks. Being the sharpest and cleverest man in the village he certainly seemed to have everything much his own way, and had wisely begun operations by marrying the belle of the village. The Chinaman's ability for adapting himself to any circumstances and for trading successfully, even under great difficulties, is truly marvellous.[75]

Chinese traders came as single men and, even if they had wives and children, their dependants had to be left behind in the villages in China. The men abroad earned money to send back to their families. Like the Hakka miners who intermarried freely with Land Dayak women, many traders took wives from among the Ibans and other natives. The native women did not object to marrying Chinese men. For the native spouses, the lot of a trader's wife appeared attractive, especially with the deceptive appearance of acquiring the worldly goods of the trader. For the Chinese men, marriage to the Ibans and other native women took care of their biological needs while being a good way of facilitating commercial intercourse with the indigenes. Below is a lengthy account of a trader who met a tragic end before he could marry his chosen native wife:

His name was Ah Sam, and he was commonly known as Assam by the natives. He was a long lean Chinaman, with a rather pallid face, slanting eyes, and strong white teeth. He talked Dayak fluently, so fluently indeed that it was quite impossible for a Dayak to understand him when he got a little excited.... He was a hawker, that is to say, he was one of the numerous class of Chinese who practically live on the water and only return to the bazaar to renew their licenses and load up afresh with a miscellaneous and rather smelly cargo of kerosene, tobacco, salt, dried fish, plates, beads and other trifles valued by the Dayak....

For four or five years Assam had regularly worked his river in common with two or three other boats, but the further Dayak houses were his particular beat, no other boat venturing so far beyond the reach of the tide. So Assam would paddle up the muddy reaches, pole up the gravelly stickles, and push and tug his boat up the rocky and swift-running stream beyond until he could get no further. He was, after all these years, an accomplished waterman and it was his custom to take the steering-paddle, while his Dayak coolies [three was the complement] would handle the poles or haul at the tow-rope as the occasion demanded.

But Assam wearied of the life in the boat; he had ambitions, one of them being to become the proud possessor of a house and land, to marry a wife and to settle down to a quiet life ashore.... And so it came about that Assam confided to his *towkay* that he was about to start on his last voyage, and that he intended to bring back with him a Dayak wife whom he had already chosen, and who would be duly installed in the new house. But being of a frugal mind he had no intention of neglecting business, and it was not until he had located his boat with the usual stock-in-trade that he bethought himself of renewing his wardrobe that he might appear before the favoured one in gala attire....

One fine day behold Assam setting forth in his boat from the bazaar, his Dayak coolies paddling lustily up-river on the tide that follows fast, and after a couple of days of hard work they came to the landing place of the house where dwells the prospective bride. For a week trading goes on as usual, but at length the cargo being exhausted, Assam announces that he intends to leave the next day and in honour of the occasion produces from the hidden-most recesses of the boat sundry square bottles of remaining Chinese arrack, together with packets of cakes and a handful of tobacco. That evening everyone is high in spirits, the drums and *tawaks* beat more and more wildly ... Assam, arrayed in purple and fine linen, that is to say in the new white singlet and blue trousers, is already gesticulating, jabbering and dancing. The merriment continues far into the night, and it is late next morning that Assam and his crew make preparation for departure....

Towards midday they hear the rush and roar of the first of a series of small cataracts, and presently, on turning the corner, come in sight of the stretch of broken water with rocks lining the banks on either side. Not a difficult passage, but one that requires a steady hand and cool nerve....

What happens now is inexplicable. It may have been a moment of mental aberration, it may have been that the rather copious overnight libations had affected Assam's nerve, but in a moment the heavy boat has swerved and is broad-side on to the current, in another moment she is rolling over the brink of the fall. The Dayaks jump clear and the bride slips not [sic] over the side, but Assam loses his balance and slides helplessly head first over the stern. The boat disappears at the foot of the fall ... Assam is being dragged up the bank ... he is dead.

The three Dayaks turn their attention to the boat which has to be righted and overhauled, and towards evening they carry Assam down the deep bank, deposit him in the boat together with such articles as have been recovered from the river, and push out into the swift current.

So Assam returns for the last time to the bazaar, and the next morning is hurried and jolted along the paper-strewn path to the over-grown burying-ground.[76]

Besides revealing the lack of obstacles to intermarriage between the Chinese and the Ibans, this story tells of the ordinary working life of a boat hawker and of his hopes and dreams, how he conducts trade with the native people, the stock of goods in trade, the material culture of the Ibans, and the dangers of riverine travel.

Though mixed marriages were prevalent, there were official obstacles which prevented mixed communities from emerging and disallowed changes in ethnic identities. The Brooke authorities were particularly wary of Chinese traders who wished to go native and become 'Dayak' by living in longhouses.[77] Chinese men with native wives who wished to reside in longhouses were not allowed to, as the Brooke officials believed that the traders who married indigenous wives were looking for an excuse to set up shop among the natives. The native women were thus forced to follow the traders and live in the bazaar shophouses or boats.[78] As for the children of racially mixed marriages, they had to be classified as either Chinese or Dayak; the Brooke officials refused to accept them as Sino-Dayak.

The adjustment of the Iban women to bazaar life and Chinese society was not without difficulty. The world of longhouses and an economic and social life centred around swidden farming was very different from shophouse life with chores like loading and unloading goods, and buying and selling, often with long days and nights of enforced idleness if no trading transactions took place. Bazaar life certainly lacked the gaiety and spontaneity of longhouse life on the *ruai*[79] and would have been dull in comparison. Without the benefit of statistics, it is not possible to say how stable the mixed marriages were.

There must have been some element of instability, judging by the existence of Brooke legislation on broken marriages and the custody of children. Some broken marriages were probably caused by Chinese traders returning to China and leaving behind their native spouses. The native women may also have found it difficult to adjust to bazaar life. In any case, failed marriages raised the problem of custody of the children. The official attitude towards child custody underwent a marked change during the reign of the second Rajah. In 1866, the government decreed that the offspring of a Chinese who married an Iban woman belonged to the father. If the mother returned to the longhouse, the father had to pay her compensation of 10 *real* ($7.20) per child. However, he could not take the child out of the country without the permission of the government. Twenty-two years later in 1888, under a new order, native women would automatically gain custody of the children if marriages broke up.[80]

Without census or marriage statistics, it is not possible to estimate the extent of intermarriages, and the accounts of observers have to be relied upon. Charles Brooke, on a visit to Simanggang in 1907, noted 'a large number of half-Chinese, half-Dayak children, cleanly kept, with fair skins and a good physique'.[81] At the down-river bazaars, despite the mixed origins of the trading communities, the cultural orientation and language spoken at home was Chinese—invariably, the Hokkien or Teochiu dialect. But in remote bazaars like Lubok Antu, the traders did not speak Chinese and they did not keep account books because they did not know how to write the ideographic characters.[82]

The Brookes were anxious to regulate the activities and movements of *ulu* Chinese traders, but with only partial success, thus enabling the Chinese to enjoy their autonomy. The Chinese bazaar communities which emerged were internally divided according to dialect group and were further opposed to each other because of trading competition. As a group, the Chinese traders competed successfully with the Malays. The relative success enjoyed by the Chinese over the Malays was responsible for ethnic tension between the two communal groups. However, the Chinese traders were accommodating towards their native clients—the Ibans, Kayans, and Kenyahs—and quite often, these social bonds were reinforced by intermarriage between Chinese traders and native women.

1. Cited in Robert Pringle, *Rajahs and Rebels* (London: Macmillan, 1970), p. 297.
2. 'Government v. Chau Soon', 5 July 1882, Simanggang CCB, Vol. 2, p. 1, SA.
3. 'Government v. Ah Toon', 15 June 1878, Sibu CCB, Vol. 3, p. 96, SA.
4. Ibid.
5. 'Government v. Cho Soon', February–December 1914, Simanggang CCB, Vol. 25, pp. 70–1, SA.
6. 'Government v. Banta, Tuai Rumah', December 1920–March 1922, Simanggang CCB, p. 75, SA.
7. 'Government v. Ah Kang', 25 May 1920, Sibu CCB, p. 212, SA.
8. In one court case, an Iban woman was warned against allowing her Chinese husband to live with her in her longhouse:
'Maoh, *tuai rumah*, states that defendant has been at his house, but he always lived in his trading boat at the *pangkalan* of the house. Defendant was collecting his debts owing in the neighbouring houses. Less than a month ago, defendant sold his boat to Dumas and only remained one night longer in the house, leaving the next day and coming down to Simanggang. When questioned, defendant stated there is another Chinese living near him at Eng Kramat, called Ah Haw, but he is in a *langkau* [hut], and not in a Dayak house. There does not seem to be any evidence against defendant. He is warned, and told he must not on any account live in a Dayak house, and his wife, if she still wants to follow him must not take him to live with her in the Dayak house.' ('Government v. Ah Fatt', 23 April 1914, Simanggang CCB, Vol. 25, p. 130, SA.)
9. 'Government v. Sui Hong', 31 May 1883, Simanggang CCB, Vol. 2, p. 104, SA.
10. In a court case, a Chinese trader was warned not to stay longer than 14 days at

UP-RIVER TRADING AND SOCIAL RELATIONS   139

a *pangkalan* (boat jetty). See 'Government v. Ah Law of Pusa', Simanggang CCB, 1 July 1913, Vol. 24, p. 29, SA. Also see *SG*, 1 April 1927, p. 93.

11. In 1919 in Simanggang, when Lim Kee Seng was asked why he did not bring his trading boat down to renew his licence, he replied that 'when the river was in fresh (flood), his trading boat got left high and dry'. 'Government v. Lim Kee Seng', 1 May 1919, Simanggang CCB, p. 21, SA.

12. *SG*, 1 December 1877, p. 198.

13. 'Government v. Kau Swee', 13 August 1906, Baram CCB, Vol. 10, pp. 228–9, SA.

14. 'Government v. Khai Tio', 9 October 1920, Baram CCB, Vol. 10, p. 142, SA.

15. 1 December 1919, Baram CCB, p. 41, SA.

16. Ibid.

17. 27 November 1920, Baram CCB, Vol. 10, p. 172, SA.

18. 'Government v. Ung Wang', 21 June 1980, Simanggang CCB, Vol. 4, p. 56, SA.

19. See note 10.

20. The preamble to Order No. XI, 1900, stated:
'I [Charles Brooke] hereby direct that on and after the first day of July 1900, all persons hawking goods whatsoever for sale or barter [except cooked food or fruit] either by land or from boats, shall take out a hawker's license.... Licenses will be issued for periods of from one month to one year on payment of a fee of 50 cents per month.' (*SG*, 2 July 1900, p. 128.)

21. 'Government v. Ah Kee', 12 August 1921, Simanggang CCB, pp. 158–9, SA.

22. Information from interview with Syn Chin Joo, 19 June 1981, Kapit.

23. Interview with Siaw Kee Siong, 26 June 1981, Kanowit.

24. Brooke officers in the Rejang River basin made infrequent upriver trips. In 1916, the officer-in-charge in Kapit failed to make his annual trip to Belaga 'due to the troubled state of the *ulu* Rejang'. (*SGG*, 17 March 1916, p. 45.)

25. *SG*, 1 April 1909, p. 76.

26. *SG*, 1 November 1939, p. 186.

27. 'Government v. Ah Piang', 1 December 1883, Simanggang CCB, Vol. 2, p. 173.

28. Letters, 30 June 1880, Vol. 1, p. 69, SA.

29. For Simanggang examples, see 'Government v. Lai', 11 December 1883, Simanggang CCB, Vol. 2, pp. 176–7, SA; 'Government v. Seng Bong', 25 April 1892, Simanggang CCB, Vol. 5, p. 61, SA. Cases from the Sibu area are 'Salt Farm v. Ah Chong', 10 August 1874, Sibu CCB, Vol. 3, p. 54, and 'Government v. Chee Kau', January 1909–March 1920, Sibu CCB, p. 133, SA.

30. 'Government v. Kasi and Chong Su Li', 11 October 1894, Betong Court Book, Vol. 2, p. 125, SA.

31. 'Government v. Ah Leng', 22 September 1923, Sibu CCB, p. 375, SA.

32. In 1921, an Engkilili trader was found to be using a yard measure which was 2 inches short. He was fined $25. 'Government v. Chop Shin Tet Shin', 21 July 1921, Simanggang CCB, Vol. 28, p. 150, SA.

33. *SG*, 1 September 1892, p. 167.

34. 'Government v. Ah Wan', 10 August 1893, Simanggang CCB, Vol. 6, p. 84, SA.

35. Information obtained from interview with Siang Tiang Choon, 5 June 1981, Sibu.

36. *SG*, 1 September 1892, p. 167.

37. *SG*, 3 May 1905, p. 128.

38. *SG*, 16 April 1912, p. 88.
39. *SG*, 3 March 1924, p. 93.
40. 'Government v. Kim Wat, Soon Guan, Yap Hin, Kim Ann and Chap Long', 30 September 1896, Baram CCB.
41. 'Government v. Chop Chiap Heng', 18 June 1914, Simanggang CCB, Vol. 25, p. 191.
42. 'Government v. Lawar, Munkong, Malisma and Rancha', 11 January 1895, Simanggang CCB, Vol. 4, p. 287.
43. Shortly after they had gone, it was discovered that the *gutta* consisted merely of a thin coating of *gutta* covering lumps of dark tree bark and rubbish. *SG*, 4 April 1906, p. 92.
44. *SG*, 16 July 1909, p. 157.
45. *SG*, 16 July 1909, p. 157, and 1 May 1924, p. 165.
46. *SG*, April 1901, p. 87.
47. In 1896, Kabong shopkeepers who did not keep their kitchens separate from their shophouses were fined. *SG*, 2 January 1896, p. 14.
48. *SG*, 1 March 1895, p. 50.
49. *SG*, 1 May 1900, p. 99.
50. *SG*, 16 September 1916, p. 200.
51. 'Government v. Choon Wai Sai, Hap Ho, Ah Loo, Swee Jak, Kua Chiu and Ah Wan', 18 December 1920, Baram CCB, Vol. 10, pp. 187-8, SA.
52. *SG*, 1 May 1900, p. 99.
53. Resident O. Lang, on a visit to Sebuyau in September 1910, exclaimed: 'I find the Chinese at Sebuyau do not pay assessment tax to the government, and it seems strange that they alone should be exempted. When asked about it, they themselves said that the government had never asked, or ordered them to pay.' (*SG*, 1 October 1910, p. 210.)
54. C. Brooke to H.H. The Rajah Muda, undated, Letters of Rajah Charles Brooke, Vol. 5, June 1898–April 1901, p. 75, SA.
55. Craig Lockard, 'The Southeast Asian Town in Historical Perspective: A Social History of Kuching, Malaysia, 1820–1970' (Ph.D. thesis, University of Wisconsin, 1973), pp. 196–231.
56. *SG*, 1 September 1908, p. 224.
57. 11 February 1888, Baram CCB, Vol. 1, pp. 145–6, SA.
58. *SG*, 3 October 1907, p. 34.
59. *SG*, 1 December 1912, p. 34.
60. *SG*, 2 February 1914, p. 27.
61. *SG*, 17 March 1916, p. 53.
62. *SG*, 1 April 1922, pp. 100–1.
63. *SG*, 16 May 1914, p. 119.
64. Lockard, 'The Southeast Asian Town in Historical Perspective', p. 209.
65. Baram Monthly Report, June 1885, p. 92, SA.
66. *SG*, 1 October 1909, p. 211.
67. *SG*, 3 February 1904, pp. 32–3.
68. *SG*, 16 May 1911, p. 91.
69. *SG*, 16 June 1917, p. 151.
70. *SG*, 16 November 1908, p. 208.
71. *SG*, 1 November 1928, p. 238.
72. *SG*, 2 October 1899, p. 208.

73. *SG*, 1 February 1893, p. 29.
74. Information obtained from interview with Syn Chin Joo, 29 June 1981, Kapit.
75. *SG*, 1 April 1885, p. 34.
76. *SG*, 1 March 1909, pp. 54-5.
77. In 1914, a Chinese trader who ceased trading and wanted to become a 'Dayak' was imprisoned six months when he was caught. See 'Government v. Cho Soon', February–December 1914, Simanggang CCB, Vol. 25, pp. 70-1, SA. See also note 5.
78. See note 8.
79. This is the veranda within the longhouse itself which leads to the *bilek*, or individual rooms of the families. It is the public area of the longhouse where people gather to talk and meet others, and where public meetings and major social occasions are held as well.
80. Cited by Pringle, op. cit., p. 295.
81. *SG*, 5 August 1907, p. 177.
82. According to information obtained from Lim Thai King, 22 April 1981, Lubok Antu.

7
The New Foochow Colonists: The Beginning of the Rubber Cash-crop Economy in the Lower Rejang, 1901–1920

THE sojourning experience of the Foochows[1] to the Rejang was part of the wider pattern of the Chinese pioneering phenomenon in Sarawak. A standing land and settlement offer of 1880, proclaimed by Charles Brooke, was negotiated with Wong Nai Siong, a Foochow, to introduce the first colony of Foochow men, women, and children to the Rejang in 1900.[2] This was the first wave of future large-scale Foochow migration that was to create a sub-ethnic enclave in the Lower Rejang. The Foochow settlers overcame the initial setbacks of padi crop failures, and in a spirit of independence, isolated themselves from other communities, catered to their own immediate needs of churches and schools, and relied upon rubber cultivation for their livelihood.

Wong Nai Siong and the Sponsorship of Foochow Migration

The well-documented story of Wong Nai Siong only needs brief retelling here. Wong was a Foochow Methodist minister, born in 1849 in the prefecture of Foochow. He took part unsuccessfully in the struggle for the reformation of the Ching Dynasty in China in 1898, fled his native Mintsing *hsien*, and journeyed to the *Nanyang*.

Wong heard of the willingness of Charles Brooke to invite Chinese agriculturists to open up the Rejang, and he subsequently visited the river basin in 1900. Satisfied with what he saw, he left for Kuching to see the Rajah. An agreement was signed with Charles Brooke on 27 July 1900, whereby 1,000 adult men and women and 300 children were to be brought into the Rejang. Wong and another partner in the venture then went back to the various Foochow *hsien* to recruit migrants, most of them from Mintsing *hsien*. Initial suspicion was encountered as it was thought that Wong was running a 'pig' slave trade network.[3] However, his standing as a Methodist minister helped

Map 8 The 'New Foochow' District

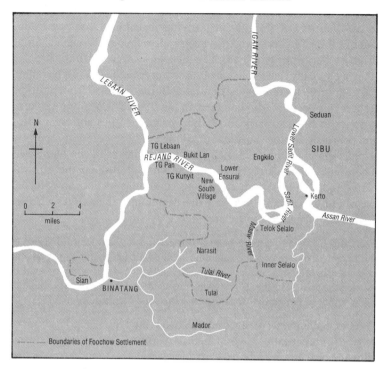

to allay the fears and anxieties of many would-be migrants.

By the end of 1900, about 500 migrants, mostly residents of Mintsing and Kutien *hsien*, had been recruited. The leaders of the *émigrés*, Lek Chiong and Ting Kwong Dou, escorted the first group of ninety-one men and women on their long journey to Sarawak, via Singapore, where nineteen of them absconded. This pioneering group of seventy-two arrived at Sibu on 28 January 1901. Three weeks later, Wong accompanied a group of approximately 500 Foochows, landing at Sibu on 16 March 1901.

What follows next are invaluable oral accounts told by various pioneers to Reverend James Hoover, an American Methodist missionary who spent thirty-two years working with the Rejang Foochows from 1903 to 1935. As a published source of oral historical recollections, Hoover's biography[4] provides a social commentary on the hopes, fears, and difficulties of the Foochows as they made their way to the Rejang River.

The Foochow prefecture was a rich and productive region, unlike other parts of Fukien and Kwangtung provinces, and this may account for the smaller numbers of Foochows who left to seek their fortunes in the *Nanyang*. A drought in 1900 was instrumental in persuading some of the Foochows, who had been approached by Wong Nai Siong to go to Sarawak, to do so. The Foochows were enthusiastic, for 'they were ready to listen to any talk of a promised land'.[5] The potential migrants had some vague notions of Borneo gleaned from returned traders in Foochow who had been to the tropical island and spoke of 'savages in Borneo, dark men who wore no clothes and who killed and beheaded for the sheer love of killing'.[6] There were also reports of strange diseases, fierce animals and deadly snakes in the *Nanyang*.[7] These tales of the strange and unknown that awaited them in the *Nanyang*, combined with calamitous conditions in Foochow, forced the would-be migrants to keenly debate amongst themselves whether or not to journey to the Rejang:

Fear held up pallid warning hands. Hunger urged them forward. Love of the home surroundings where families had dwelt for centuries argued for remaining in Mintsing, while youth and adventure pleaded that young families start on the great journey. A few even argued with idealism, that the going of a few hundred emigrants would allow more food for those remaining.[8]

Arrangements were eventually made for the second group of pioneers prepared to leave in February 1901, after some delay and allowing time for settling debts and for family feasts and formal calls. Even at the last minute, there were reservations about the trip. The emigrants gathered at the Wong family home, each able-bodied person carrying a load on the back or hanging from the bamboo *bieng-dang* slung over the shoulders. Small children were placed in baskets suspended from the *bieng-dang*. At Foochow City, they awaited the sea-going junks. Several times, during the period of waiting, they were called together at the Church of Heavenly Peace for prayer services.[9]

When the sea-going junks, *Bing-hai* and *Hai-cue*, were ready to sail for the *Nanyang*, the colonists went by *sampan* to the vessels and clambered on to the decks:

The junks had stanch timbers of the hulls rounded on the outside. High on the prows were great circles of wood, painted to resemble eyes, by means of which the junk was supposed to find its way across the seas. Substantial railings hemmed in the small deck, on the aft hold of which was erected a rough but strong housing. It was a cabin. Into the entirely inadequate cabins and down into the darker, foul-smelling holds poured the human stream, men, women and children. Each had its own bedding. Each had brought such clothes and personal belongings as were possessed and could be carried.[10]

THE NEW FOOCHOW COLONISTS: 1901–1920   145

Conditions on board the junks were unsatisfactory and rough. When the waves washed the decks, the passengers returned to the cabin, wrapped in their bedding which served as blankets. They could not stand the stinging cold winds and waves. After six days and nights, the disheartened travellers reached Hong Kong. Many of them were ill with dysentery, and most of them were seasick and just felt miserable. In Hong Kong, the migrants met Bishop Warne of the Methodist Church; encouraged by the Foochows' intended journey to the Rejang, he agreed to accompany them. As the vessels approached Singapore, the pioneers had their first experience of the heat and humidity of the tropics:

... And always the heat grew more oppressive, the glare of the sun more unbearable. The eyes of the emigrants, many of them weakened by the prevalent trachoma, smarted and turned blood-red. Again they sought shelter in the cabin, but now from the blistering biting heat of the sun, instead of the cold winds already forgotten.[11]

With the rough sea journey from Singapore behind them, the migrants began the start of a tough new life, which required their adjusting to a different climate and physical environment in the Rejang. They were to live on a tropical frontier just recently settled by pioneering Ibans, and this too added to their difficulties and problems.

Pioneering Conditions in the Lower Rejang

The first seventy-two pioneers who arrived in January 1901 were divided into two groups according to their origins. Foochows from Kutien *hsien* were settled at Ensurai, while those from Mintsing *hsien* were sent to Sungei Merah or Sungei Seduan. Two temporary huts were built by the government at Ensurai, while six were constructed at Sungei Seduan. The second group of about 500 colonists were located according to native-place origin: 200 Kutien Foochows were sent to Ensurai, and the remaining Mintsing people were put up at Seduan. At the end of 1901, Wong Nai Siong returned to the various Foochow *hsien* to recruit more people. Another 500 Foochows were brought to Sibu on 7 June 1902.[12]

The government-built cabins were a staging post for the new settlers. A Foochow pioneer who subsequently wrote an account of his pioneering days was Lin Wen Tsung. According to Lin, the six cottages at Sungei Seduan in fact formed a single long building, partitioned by *attap* leaves and with floors built of small tree branches lashed together. Each of the six huts was about 16 feet wide and

30 feet long.¹³ At the back of the 'longhouse' was a shared kitchen with two rows of seven or eight big Chinese cooking pans.¹⁴

Though the housing was comfortable enough, the hot and humid weather was a problem for the settlers. According to oral reminiscences, no matter how fresh the meat and fish were, they had to be cooked and eaten on the very day they were obtained.¹⁵ The pioneers had to bathe three or four times a day. They were fearful of taking afternoon naps, afraid that they might contract fever, or die in their sleep. A popular refrain went like this: 'Today, you go and bury someone, tomorrow, you may be buried by somebody else.'¹⁶ This was a reference to the prevalence of illness and death among the Foochow pioneers. By the end of 1901, some twenty settlers had fallen ill and died.¹⁷ By then, they had begun to move out of the government longhouses, building their own huts, while those without accommodation slept on grass mats spread out under the sheltering trees. Living in the open was not too uncomfortable in good weather, but when the rain fell, the pioneers had to sit up and drape the mats over themselves in tent-like fashion. Food was cooked on open fires, and on rainy days it was not possible to light fires. The move out of the staging posts was the beginning of the shattering of the hopes and dreams of many of the colonists.

Despite the initial hardship of adjustment, the colonists encountered success in vegetable growing, and the *Sarawak Gazette* painted an optimistic picture of agricultural conditions: 'More than fifty acres of land at the Seduan settlement are now neatly laid out and planted as market gardens, containing a great diversity of vegetables, yams, plants and nuts; a still greater acreage has been reclaimed by their fellow colonists at Ensurai.'¹⁸

As a result of over-production and the lack of a market, there was a glut of vegetables and sweet potatoes in 1901. The *sinkheh* were to encounter worse problems when they began planting padi, one of the objectives for the establishment of the colony. The first three years in which the Foochows experimented with rice cultivation were the toughest.

The jungle loomed vast and threatening. The sight of tattooed Ibans, with knotted hair on their heads and *parang* by their sides, was an even more frightening spectre for the would-be rice farmers. The first year of planting was disastrous. Rice-growing methods, successful back in Foochow country, did not work in Sarawak. In the ricefields in Foochow, all weeds, pests, and stones could be removed. In Sarawak, however, it was not possible to do this; moreover, the soil was porous and full of roots, and irrigation was difficult. With the onslaught of rain, the rice stalks and seedlings were washed out and there was no harvest. Wong Nai Siong as 'proprietor' of the colony, was obliged to feed the settlers out of funds borrowed from the Rajah.[19]

The following year, 1902, proved to be just as disastrous as the first. The Foochows tried planting two crops. The first crop failed because it was sown early in the season—too early, as it turned out—to make way for the second. The second crop brought no harvest because of bird pests.[20] Wong Nai Siong was convinced by then that the soil and climate were not suitable for traditional wet-rice cultivation. After a visit to the Malay Peninsula and Singapore to study rice planting and to look at possible alternative crops, he returned to the Rejang and mistakenly advised the pioneers to plant sugar-cane. The crop grew well, but there were no mills to process the cane, and there was no market.[21] It was like the sweet potato glut of 1901 all over again. Once more, Wong Nai Siong, as sponsor and leader of the Foochow colony, was obliged to feed the planters. The climate, the harsh physical environment, illness, and successive crop failures adversely affected the morale of the Foochows.

It was reported that some had died, others had 'vanished' in the jungle, floods had taken some, and a few had been able to stow away on steamers or junks trying to escape from this 'tropical paradise'.[22]

After the repeated failure of planting techniques imported from the Foochow homeland, the colonists in the third year tried the Iban swidden method of slashing and burning and then planting, but to no avail. By the end of the third year, they wanted to abandon the settlement and return home; Wong Nai Siong had to cajole and coerce them to stay as he was in debt to the Rajah and had to find ways of repaying the money. Of the 1,000 or so original inhabitants, it has been variously estimated that there were only between 500[23] and 673[24] remaining.

In order to repay the money owing to the Rajah, Wong Nai Siong proposed to tax the farmers a percentage of their produce. This proposal aroused heated protests from and resentment amongst the impoverished Foochows. After four years of disappointment and suffering, the pioneers were so discouraged that they refused to undertake any cultivation, and lived almost entirely at the expense of the sponsor. Unable to repay the loan to the Rajah, Wong was left with two choices: either account for how the loan of $40,000 was spent, or leave the country.[25]

Wong Nai Siong had managed the colony badly. When he started the venture, he had little, if any, knowledge of agriculture. He did not have an agricultural background, being a Methodist minister with idealistic visions of creating a prosperous colony, along the lines of the Plymouth Pilgrims establishing themselves in America and from whose history he had received inspiration. Wong did not foresee the practical difficulties in planting, in the adjustment the pioneers would have to make to their new environment, and he was not capable of tackling the problems which arose. Further, he handled the finances badly. About half of the loan money was spent on boat fares from Foochow and on farming tools and equipment unsuitable for use in Sarawak. He had also lost another $4,000 when one of his helpers absconded with the money on the pretext of recruiting migrants from Foochow prefecture.[26] But Wong Nai Siong cannot be held entirely responsible for the failure of this first phase of Foochow migration to Sarawak. The pioneers themselves were markedly insular, and their insularity did not help them either to adjust to their new environment or to receive assistance from the Ibans. When Wong Nai Siong finally left Sibu in 1904, the task of managing the floundering colony fell to an American Methodist missionary, James Hoover.

The new leader of the Foochows first arrived in Kuching in 1903, and received 'a courteous but casual welcome from government officials'.[27] The officials had reservations on the value of missionary work among Foochow migrants, about whom complaints had been received over land quarrels with the Ibans. The Brooke administration had always considered itself the protector of native interests, and was therefore unhappy over the land disputes in the Rejang. Hoover, though not unaware of the disputes, was more concerned with the management and well-being of the Foochow colony. He was unhappy with the old system of a 'proprietor' looking after the welfare of his people. The missionary wanted to see a self-sustaining, self-reliant community. His efforts—as a source of inspiration and innovative ideas, as a leader, and as a mediator between government and people— were to prove a catalyst to the Foochows. Summing up the reasons for Foochow failure and the rapid decline in Foochow numbers, he said: 'Sickness, loneliness, failure of crops and wild men [Ibans] took their toll.'[28]

Over the next several years, conditions improved dramatically for the Foochows. When Hoover absolved himself of the responsibility of feeding the colonists, they had either to plant successfully or starve. As it turned out, and it is not known how or why, the Foochows reaped good harvests after the initial years of successive rice failures. Perhaps it was the 'choice' between survival or famine which spurred the settlers into planting and caring for their padi crop properly. They no longer had a sponsor to feed them in time of need. Growing enough food for themselves did not satisfy the settlers when they saw that they could earn extra income by planting other crops.

There was a quick and judicious switch to the cash crops of pepper and rubber. For example, in 1906, 40,000 pepper vines were planted and 2,000 pioneer rubber seedlings were sown by two individuals, Wong Ging Ho and Kuok Yew Dew, and 'four thousand more plants and three thousand seeds were expected by the end of March'.[29] Hoover saw in rubber a remunerative cash crop which would solve the financial woes of the colony, and enable the Foochows to become financially independent. Many Foochows were of similar opinion and needed little encouragement to turn to rubber planting. In fact, they had to be constantly reminded not to abandon altogether rice cultivation, the original goal for establishing the settlement.

With their new-found success, the Foochows still faced the nagging problem of repaying the loans taken from the Brooke administration by Wong Nai Siong. The settlers petitioned the Rajah to waive the payment of their outstanding debts on the grounds that they were unable to pay. Rajah Charles Brooke agreed to this request on condition that the pioneers pay an annual rental of $1.00 per acre and an annual quit rent of $0.50 per year on the land they occupied.[30] The cancellation of the loans pleased the Foochows, who now felt they were better able to consolidate their financial position and entitlement to the land. The issuing of land titles gave them a sense of permanency, for they now owned the land they tilled. The social structure of the Foochows, reinforced by their churches and schools, was instrumental in helping the pioneers adjust to their new environment during their initial years.

Social Organization

The Methodist Church

There was a close relationship between the Methodist Church and the growth and consolidation of the Foochow colony. The majority of the Foochows were Methodist. Non-Methodist Foochows were encouraged to embrace Methodism.[31]

Wong Nai Siong, the Foochow sponsor and subsequent 'proprietor' of the colony, was an enthusiastic missionary and preacher and, from the pulpit of the Methodist Church in Mintsing *hsien*, had first introduced the 'tropical paradise' to prospective migrants.[32] Churches in the Foochow prefecture were the communication points for news about the 'New Foochow' in the Rejang. When Dr West of the Methodist Mission visited the Foochows in November 1902, he immediately drew up plans for the establishment of a mission headquarters at Seduan.[33]

James Hoover, who was already doing missionary work in Malaya, volunteered to go to Borneo in 1903. He was a keen worker and was anxious to see the Foochows settled into their new home. The missionary believed his objectives in the Lower Rejang to be not only spiritual, but temporal as well. Hoover saw no contradiction between spiritual and material requirements.[34] Neither for that matter did the Foochows, for whom the primary purpose of journeying to the Rejang had been economic well-being, and the more staunch in their faith they would become if the Methodist Church helped to improve their welfare. As conditions improved for the Foochow settlers, churches were established in quick succession in the various settlements. The Sing Chio Ang Church at Sungei Merah was one of the first chapels to be built, in 1903. As the Foochows spread out from the pioneer settlements of Seduan and Ensurai to three other villages, South Village, Upper Village, and Lower Village, they built wooden chapels at each of these three places. The villages contributed labour and donated $2 for every $1 of missionary funds provided.[35]

The Church impinged directly on the lives of the Foochows in several ways. The leadership of the community was linked to the Methodist Church in that church leaders were often prominent figures of the community. The Foochows who migrated to the Rejang saw the Church as a rallying point and as a refuge. They needed the Church as a place of solace and comfort in a land which was often frightening to them and whose Iban inhabitants they did not understand. The Church, in the pioneering days of the colony and in the absence of other community associations (which were subsequently formed at a later period), was one of the pillars of Foochow social life and community organization. Churches and schools usually occupied the same premises, or stood adjacent to one another. In 1904, Ethel Mary Young, wife of James Hoover, writing in her diary, mentioned how churches and schools were functionally related: 'Our first home was in one end of a building that was used as a school on week-days and a meeting-house on Sundays. We had two rooms each 11 by 11 feet, which were sitting room, dining room and bedroom all in one—or two.'[36]

By 1905, the Hoovers were better housed. They had a building with four rooms—one for themselves, the second for the boys' school, the third for the girls' school, and the last, a chapel and quarters for the Foochow preacher.[37] In 1917, a central church was erected at a cost of $1,200, but it became inadequate to meet the needs of an expanding population and, in 1925, a larger church was built.[38] The Methodist Church, according to local informant Ling Kai Cheng, began proselytizing in earnest only after 1920.[39] This could have been because the Foochow settlements were more established by then. Churches and schools expanded in the third decade of the twentieth century, in the wake of the rubber boom which attracted a constant flow of Foochows to the Rejang.

Education

From the start, the Foochows placed a high premium on the value of education for their children. Even the scattered Foochow villages along the banks of the Rejang River each had independent schools.

In 1903, Hoover set about establishing an exclusive Foochow school with an enrolment of thirty boys, half of whom were to be boarders. The school had two classrooms, one of which doubled as a dormitory and the other as a church.[40] Officially called the Kwang Ang Primary School, the school taught in both English and Chinese. It was to be the first of a number of similar schools, all of them one-teacher establishments.[41] Hoover's idea was that schools like Kwang Ang should not be supported by the government, the adult Foochow community, or the Church, but rather, they should find their own means of support. As a corollary, the notion of independence and hard work was inculcated early in the schoolchildren. Seeing rubber as a valuable cash crop and a means of financial support, Hoover actively encouraged the growing of rubber trees in school compounds. Around each school was a garden and a small rubber plantation, the income from which helped to meet expenses.[42] This method of financing schools began in 1905. The pupils tapped the rubber trees early in the morning, and after this activity was completed, school lessons would begin at 11 a.m and go on until 5 p.m.[43]

In 1905, the Kwang Ang Primary School was shifted nearer to Sibu, in the vicinity of the present wharf area. Mrs Hoover taught there as well, concentrating on female students. Eventually, a separate Yit Ing Girls' Primary School was founded.[44] Mrs Hoover, an active and dedicated teacher, taught in the boys' school (the Kwang Ang) in the morning, and at the girls' school in the afternoon. She also found time to conduct Bible classes and to teach in the homes of

Foochow girls who were too old to be seen out of their houses, or whose bound feet made it impossible for them to walk to school.[45] Three years later, in 1908, a third Anglo-Chinese school was built at the site of present-day Island Road in Sibu. In 1913, this school was converted into an Industrial Agricultural School after Hoover had obtained a grant of $500 from a benefactor in Kansas. Rajah Charles Brooke allocated 250 acres of land for the school, which was eventually located at Bukit Lan, 10 miles down-river from Sibu, and promised an annual grant of $500 to defray expenses.[46] By 1917, two other schools had been built, the Kwang Nang Primary School at South Village, and the Chung Hin Primary School at the New Henghua Village.[47] The Methodist Mission contributed to the cost of constructing both schools. It also paid the salaries of the teachers and covered the general maintenance costs of the schools. Steps were taken in December 1916 to build the first secondary school and it was decided that a temporary school should be set up within the residential premises of an individual named Ting Chen Ern. Hoover's permission was obtained to move the Industrial Agricultural School at Bukit Lan, its teachers, students and equipment to the temporary school, subsequently named Kwang Hua Secondary School. It had an initial enrolment of 105 male and female students.[48]

The growth of schools up to the end of the second decade of the settlement's existence was generally slow, because the schools were meant to be self-supporting and received no government help. The rapid expansion of the rubber economy in the 1920s, together with an increasing population, accelerated the growth of schools in the next two decades. The schools, besides fostering and symbolizing the independent spirit of the Foochows, were important as a transmitter of Chinese education and culture, particularly as they catered exclusively for Foochow students. The role of schools in helping the Foochows to reinforce their sense of being Chinese and Foochow, to the exclusion of others, was a crucial factor that enabled the Foochows to close in on themselves as a community. Other changes, such as the economic attraction of rubber, were equally important in fostering a closed community.

The Rubber Economy

If there was any single factor other than the Methodist Church or the schools that could be considered a dominant influence on the Foochows, it was rubber. Government policies such as exclusive

Foochow land reserves, the liberal issue of land titles, and a *laissez-faire* attitude to Foochows clearing land without proper cadastral procedures were all encouraging to prospective settlers. The legendary work ethic and the drive to be self-sufficient and to save money were social characteristics of the Foochows which helped them take advantage of the liberal Brooke policies regarding rubber cash-cropping.

In August 1909, the Rajah agreed to set aside 12 miles square of the river bank on both sides of the Rejang below Sibu as the Foochow concession, and no Ibans were allowed on the reserve. Tacit encouragement was given to rubber planting as no rent was charged on land taken for its cultivation. In place of rent, a duty of 5 per cent was imposed on rubber exported after 1 January 1910.[49] The reserve was granted for two reasons: to encourage the Foochows to cultivate rubber, and to prevent any further ethnic conflict over land. A 1914 Land Order provided for survey fees of $1.00 per acre. There was nothing novel in Brooke policies concentrating the Chinese within certain areas. The Brookes had already encouraged and then forced Chinese traders to build bazaars so that they would not be scattered amongst the natives. As a result, the Brookes had unwittingly created a Chinese identity in towns and bazaars. However, the effect of Brooke immigration policies with respect to the Foochows deliberately created the opposite effect—that of a rural, agricultural orientation.

The first rubber trees in the Foochow concession were tapped around 1910 and 1911, and the planters were thrilled when rubber prices fetched $700 per *pikul*. A legend was built up around rubber trees which were described as 'trees in which you shake money'.[50] By the end of 1910, more and more Foochows were turning to rubber at the expense of padi. Resident Baring-Gould lamented: 'When considering the large area made over to the Foochows last year [1909], it is disappointing that a larger area was still not laid out in padi.'[51]

The ease with which the Foochows could now apply for ownership of land within the framework of the 1909 concession, a constant stream of migrants from China, and high rubber prices, all combined to result in a spreading of Foochow settlements along the banks of the Rejang. The second decade of the twentieth century saw the peak years for registered Foochow arrivals, with 601 migrants arriving in 1911, 846 in 1913, and 1,227 in 1916. This is illustrated in Table 7.1. Rubber prices fluctuated widely on the world market, for example, declining from an average price of 8s. 9d. per pound in London in 1910 to a low of 2s. 0.75d. per pound in 1919. But this variation in rubber prices did not affect the constant flow of migrants to the Lower Rejang as Table 7.1 suggests.

TABLE 7.1
Foochow Arrivals, Land Titles, and Rubber Prices,
1901–1920

Year	Rubber Prices (per pound)		Foochow Arrivals				Land Titles
	s.	d.	Men	Women	Children	Total	
1901	n.a.		n.a.	n.a.	n.a.	500	n.a.
1902	n.a.		n.a.	n.a.	n.a.	500	n.a.
1909	7	1.00	n.a.	n.a.	n.a.	n.a.	24
1910	8	9.00	n.a.	n.a.	n.a.	386	127
1911	5	5.50	416	80	105	601	311
1912	4	9.00	117	26	30	173	416
1913	3	0.25	674	91	81	846	109
1914	2	3.50	n.a.	n.a.	n.a.	n.a.	n.a.
1915	2	6.00	68	15	n.a.	83	181
1916	2	10.25	1,080	147	n.a.	1,227	154
1917	2	9.75	508	140	188	836	372
1918	2	2.75	80	9	16	105	456
1919	2	0.75	155	43	63	261	682

Sources: *SG*, 1901–20; *SGG*, 1911–20; Land Office files; and Rubber Growers Association, *Statistics Relating to the Rubber Industry* (London: 1928), Table 14, p. 20.
n.a. = not available

For a recorded total number of 4,035 arrivals (men and women) from 1910 to 1919, 2,808 land titles were issued. This would suggest that some Foochows did not apply for or take out land titles. This may be explained in part by the fact that some Foochows worked as wage-labourers in rubber gardens owned by others. A second reason could be that land titles issued were an inaccurate index of land actually cleared, as settlers often opened up land without ownership papers. Finally, the understaffed Land Office, according to its own reports, was unable to keep pace with the cadastral work and was often unaware of land having been cleared for cultivation by the Foochows. Moreover, the Land Office had no properly developed guide-lines on land policy, especially concerning land owned through native customary law.

Preparing land for planting rubber trees became central to the Foochows, and various pioneers have related that rubber was the *raison d'être* for their migration to Sarawak. The new arrivals claimed that planting rubber was the only means of earning a livelihood. There were other crops which they could plant, such as padi or pepper, but they were never as attractive. Rubber tapping was a skill which most Foochows became acquainted with even at an early age, and men, women, and children learned the art without exception.[52] Planting rubber was a long-term investment, the trees taking seven to eight years to mature. New arrivals would usually work as wage-labourers[53] or on a share-crop basis.[54] The dream of most settlers was to become landowners. A typical rubber garden could range from 2 to 10 acres.[55] Most of the land would be taken up by rubber trees with side patches of padi and vegetables, grown as supplementary crops to feed the planter and his family while they were waiting for the trees to mature and at times when tapping was not possible because of bad weather.

A typical tapping routine might begin at about 4 a.m. or 5 a.m., and the work of collecting and coagulating the rubber latex would not be completed until 8 a.m. or 9 a.m. Women assumed heavy responsibilities in both garden and home.[56] After the day's tapping was finished, the women returned home and performed routine household chores like laundry, cooking, and looking after small children. Work on the vegetable plots and rice-fields was undertaken in the afternoons. Apart from the occasional clearing of weeds and small plants, rubber smallholdings needed little maintenance, once established. Estimates of rubber yields varied from 1 *pikul* to 2 *pikul* per month per acre, calculated on an average of twenty working days a month, and allowing for disruption of work due to rain.

The returns from rubber tapping and trading were unpredictable on account of several factors. First, the rainy season from October to March restricted tapping considerably. Second and more importantly was the fluctuation in prices, with local conditions of boom and slump tied directly to price changes on the world market. In 1919, the local price of rubber shot up to $100 per *pikul*, only to plummet to $15 in 1921. Whenever rubber prices fell, the Foochow colonists wisely concentrated their efforts on padi and vegetable growing.

Within two years of the creation of the Foochow reserve in 1909, the area of land under rubber had doubled, with many more applications unprocessed. Indeed, the Land Office was much understaffed and was unable to deal with the cadastral work: in 1911 there was only one qualified surveyor with an assistant.[57] When Baring-Gould, the Resident, visited the Foochow villages in 1911, he noted that the settlers were forsaking padi and pepper for rubber:

The acreage under rubber is extending rapidly and the pepper gardens are well cared for but are not being enlarged. About the same area of land has been planted with padi as in previous years.... It does not seem that these people will ever plant more padi than is necessary for their own consumption as they find other products will give them a more valuable return for their labour.[58]

The Brooke administration attempted to persuade the Foochows not to abandon the cultivation of padi:

> In 1915, the American Episcopal Mission reported that the Foochow colonists did not intend to plant padi this farming season but would put all their time to the extension and improvement of their rubber plantations, from which they are obtaining a handsome income. I [Baring-Gould] thought it advisable to issue a notice to the colonists pointing out that in the event of a shortage of imported rice next year [1916] they would be in difficult circumstances. The Mission gave great assistance in impressing on the colonists the advisability of not suspending padi cultivation.[59]

To put pressure on the Foochows to plant padi, a government decree was issued in 1918, prohibiting the planting of rubber and other cash crops on land alienated specifically for padi cultivation.[60] While rubber prices were high, warnings like this were ignored as the planters were concerned only with maximizing their returns from the land,[61] and the temptation to convert rice-fields over to rubber gardens was very great indeed.

By 1920, Sarawak was becoming increasingly dependent on imported rice. The Brookes' fear of the consequences[62] of this deficit was realized when rubber prices fell to a low $0.25 a *kati* in 1920. Rice locally grown for local consumption was insufficient. Many Foochows, whose hopes rested on continually high prices for rubber, were badly affected. A number of them sold their gardens and returned to China.[63]

To help alleviate the rice crisis, H. L. Owen, the Acting Resident, obtained a piece of Iban farming land near Sungei Pan, below the Foochow concession, and had it subdivided into 3-acre plots to be rented out at nominal amounts to the Foochows specifically for padi cultivation.[64] In 1920, to allow for rice cultivation elsewhere in the Rejang without undue delay, 'occupation tickets'—giving rights of ownership without survey work being first undertaken—were issued for the first time.[65] This administrative action, designed to boost rice production and to ease the Land Office of part of its workload, only helped to accentuate difficulties and confusion in land matters over the course of the next several years.

Rubber prices had risen again by the end of 1923, causing disruption to Foochow planting. Rice-fields which had been earnestly tended during the periods of rice shortages in the early 1920s were hastily abandoned. This time, rice had to be imported from Limbang.[66] The attitude of the Foochows was that 'they [can] make so much money out of rubber that at anything over 50 cents a kati they (can) even afford to pay high prices for padi'.[67]

1. Apart from Sarawak, Perak in Peninsular Malaysia and Singapore are the other places where there are concentrations of Foochows in the *Nanyang*.
2. In his autobiography, Charles Brooke wrote:
'If there ever should be a large influx of Chinese agriculturists this would be a most desirable locality for them. It is a country which they would develop by opening out a communication by road to other places. Besides, the great advantage consists in the river being navigable for large vessels, and in the event of Chinamen becoming troublesome and rebellious, an attack could easily be made upon them, and communication cut off from the coast on which the supply of opium and other articles would depend. This would effectually curb their unruliness.' (Charles Brooke, *Ten Years in Sarawak*, Vol. 2 (London: Tinsley, 1866), p. 252.)
3. See W. L. Blythe, 'Historical Sketch of Chinese Labour in Malaya', *JMBRAS*, Vol. 20, No. 1 (1947), pp. 64–114, for more details on indentured labour recruitment practices in nineteenth-century Malaya.
4. Frank T. Cartwright, *Tuan Hoover of Borneo* (New York: The Abingdon Press, 1938), p. 55.
5. Ibid., p. 56.
6. Ibid., p. 57.
7. Ibid.
8. Ibid.
9. Ibid., pp. 58–9.
10. Ibid., p. 60.
11. Ibid., p. 66.
12. Various estimates have been made of the numbers arriving as part of these pioneering groups. Lau Tzy Cheng in *Wong Nai Siong and the New Foochow* (Singapore: Nanyang Institute, 1979), p. 14 (text in Chinese), gave a total of 1,118 arrivals; C. V. Brooke in *SG*, 4 June 1904, pp. 124–5, recorded 997; and Cartwright, op. cit., p. 97, estimated about 1,072 newcomers.
13. Lin Wen Tsung, 'The First Ten Years of the New Foochow Colony in Sarawak', *The Sarawak Teacher*, Special History Edition, Vol. 2, No. 2 (1966), p. 13.
14. Lau, *Wong Nai Siong and the New Foochow*, p. 17.
15. Ibid., p. 19.
16. Ibid.
17. *SG*, 1 November 1901, p. 220.
18. *SG*, 1 July 1901, pp. 142–3.
19. Cartwright, op. cit., p. 54.
20. *SG*, 2 June 1919, p. 140.
21. Cartwright, op. cit., p. 47.
22. Ibid., p. 97.
23. Ibid.

24. *SG*, 28 April 1904, pp. 124–5.
25. Lin, op. cit., p. 14.
26. Lau, *Wong Nai Siong and the New Foochow*, p. 14.
27. Cartwright, op. cit., p. 47.
28. Ibid., p. 84.
29. *SG*, 4 April 1906, p. 64.
30. Cartwright, op. cit., p. 67.
31. Ibid., p. 57.
32. Ibid.
33. *SG*, 3 February 1903, p. 35.
34. Material well-being, however, did not extend to the pastimes of smoking opium and gambling, common among the Chinese elsewhere in the state. When James Hoover became 'Protector' of the Foochows, he maintained the ban on opium smoking and, in October 1922, requested the authorities to raid an opium den at Ensurai. Two men, Chung Ah Ngee and Hu Liang Kong, were caught and fined $50 and $75 respectively, and it was pointed out to them that 'under agreement with the government, the Foochows were not to smoke opium without the headman's consent and supply'. 'Government v. Chung Ah Ngee and Hu Liang Kong', 9 October 1922, Sibu CCB, pp. 65–6, SA.
35. Cartwright, op. cit., pp. 35–6.
36. Nathalie Means, *Malaysian Mosaic: A Story of Fifty Years of Methodism* (Singapore: Methodist Book Room, 1935), p. 119.
37. Ibid.
38. Cartwright, op. cit., p. 97.
39. Information obtained from Ling Kai Cheng, who first arrived in Sibu in 1913 with a group of forty Henghuas. Personal interview with Ling Kai Cheng, 5 June 1981, Sibu.
40. Cartwright, op. cit., p. 82.
41. Teachers were recruited mostly from China. One such teacher was Law Hieng Ing, who was engaged in 1926 to teach in the Kwang Nang School at Sungei Maaw which had about seventy pupils. Interview with Law Hieng Ing, 15 May 1981, Sibu.
42. Cartwright, op. cit., p. 106.
43. Personal interview with Teng Tung Hsin, 21 June 1981, Binatang.
44. Lau, *Wong Nai Siong and the New Foochow*, p. 38.
45. Means, op. cit., p. 120.
46. Cartwright, op. cit., p. 97.
47. *SG*, 16 October 1917, pp. 261–2.
48. Lau, *Wong Nai Siong and the New Foochow*, p. 39.
49. *SG*, 16 August 1909, p. 180; and 1 October 1909, p. 210.
50. Lau, *Wong Nai Siong and the New Foochow*, p. 36.
51. *SG*, 16 March 1910, p. 68.
52. Information from Wong Kwong Yu, 30 May 1981, Sibu.
53. One such person was Ling Chu King. Interview with Ling Chu King, 21 May 1981, Sibu.
54. Madam Law Suok Moi, who arrived in 1940, worked with a relative in a 10-acre rubber garden on a shared basis. Interview with Madam Law, 19 May 1981, Sibu.
55. This deduction was made after examining the Land Office records in Sibu.
56. Information obtained from Madam Law Suok Moi of Sibu.
57. *SG*, 16 March 1911, p. 54.
58. *SG*, 1 May 1911, p. 81.

59. *SG*, 1 November 1915, p. 253.
60. *SG*, 2 January 1918, p. 5.
61. Information from Ling Chu King.
62. *SG*, 1 November 1920, p. 241.
63. *SG*, 1 March 1920, p. 62.
64. *SG*, 1 June 1921, p. 111.
65. A. F. Porter, *Land Administration in Sarawak* (Kuching: Government Printing Office, 1967), p. 47.
66. *SG*, 1 May 1924, p. 157.
67. Ibid.

8
The New Foochow Colonists: Social Relations and Economic Transformation, 1921–1941

BY the beginning of the 1920s, the scattered Foochow communities had begun to expand along the banks of the Rejang River. Foochow society was becoming more complicated, more churches and schools were being built, and formal associations were being formed to cater for particular needs. The settlers were becoming a self-sustaining community. The Foochows required more land, but both the economic aspirations and the isolation of the Foochows as a community were challenged when Ibans in Binatang opposed their acquisition of additional land.

The Spread and Expansion of Foochow Settlements

As the immigration figures after 1909 suggest (see Tables 7.1 and 8.1), once the Foochows were granted a permanent concession, there was a constant stream of incoming settlers. Rubber needed plenty of land and the concession area was considered by the settlers to be insufficient for their needs. The Foochows were not allowed to move north of Sibu because of the possibility of friction with the newly developed Cantonese settlement there.[1] The alternative was to move down-river in the direction of Binatang and Sarikei.

The large-scale settlement of Foochows in the Binatang district did not get under way until 1922, although a couple of enterprising Foochow individuals, Tiong Kung Ping and Ling Ming Lok, had been to Binatang in 1908 and 1910 respectively, on personal and business visits.[2] Even before Binatang was settled, the shortage of land felt by the Foochows was manifest in the colonists overstepping the limits of their 1909 concession.[3] The opening up of the Binatang frontier was at the behest of individual initiative: two Foochows, Yau Shao Ching and Ling Ming Lok (mentioned above), requested Hoover to apply to the Resident for permission for Foochows to settle there. This request was immediately rejected, but an intercession to the Rajah subsequently reversed the original decision. Three hundred Foochows then moved in to settle in Malow, Tung-lai, Mankong, and Krupok.⁴ The next phase of the colonists' pioneering efforts was a traumatic one as the neighbouring Ibans opposed their presence. The spread of Foochows to Sarikei was also due in large part to the efforts of enterprising individuals. In 1910, some Mintsing Foochows—Wong Ching Poh, Wong Ching Chung, Lau Yen King, and a few others⁵—had rowed a boat to a Malay kampong in Petelit, where after obtaining permission from the headman, Haji Omar, they cleared the jungle and started to plant padi, vegetables, and fruit trees.⁶ In 1914, Wong Tien Pau and a few others arrived at Lubokan, where land allocated to Cantonese pepper growers had subsequently been abandoned. Almost immediately, quarrels broke out with the Ibans over land demarcations. The Land Office acted as arbitrator, and the land was apportioned between the disputants. Shortly after, the Foochows returned to the area in larger numbers and established Lubok village. In 1910, some Foochow colonists went to Kanowit in spite of official disapproval and, in 1935, they moved further up-stream to Nanga Dap, and then to Ngemah in 1943.⁷

It has been suggested that the rapid geographical expansion of the Foochow settlements was due to liberal land laws.⁸ The initial Foochow colonists did indeed migrate to Sarawak as a direct result of government policies. The Foochow settlers who followed, came on their own initiative, tempted by the riches to be made in planting the 'money-trees' (rubber trees) in a 'tropical paradise', and influenced by pioneering kin, family, and friends into joining them in the 'New Foochow'. Once established in the Rejang, these enterprising agriculturists did not remain content for long with just a small area of land to work in and around Sibu, and soon spread out along the Rejang River and into other parts of the state.

The Land Disputes

The fundamental cause of the land disputes was the issue of how and where one could rightfully acquire land. There were a number of factors to consider: Iban customary law, the needs of the Foochows, and an official Brooke supra-structure to look after the conflicting needs and rights of the Foochows and the Ibans. Against this has to be placed the background of economic change which saw both Foochows and Ibans scrambling on to the rubber cultivation bandwagon in response to the industrial needs of the West (for the motor vehicle industry and for military use in the First World War).

The land disputes started almost immediately after the arrival of the Foochows in the Rejang.[9] When Reverend James Hoover took over the leadership of the colony, he perceptively identified the causes of the friction:

> Since the proprietor of the colony was sent off by the Rajah, the Chinese have prospered wonderfully, and are pushing out on all sides and ends. The Dayaks are beginning to see what is coming and are bracing themselves and contesting almost every new clearing. They were perfectly willing to have the Chinese among them as long as they had only a patch of potatoes that tempted the wild pigs within striking distance, but when the Chinese began to open pepper plantations and rubber gardens that threaten to take every available piece of high land in the district, they wonder what is to become of themselves.[10]

Hoover and the Resident acted as arbitrators, surveying and putting up boundary posts. Where the Foochows had accidentally cut down fruit trees, they were asked to compensate the Ibans for these trees. Where the Ibans had their rice-fields, the Foochows were kept off the land.

Charles Vyner Brooke, who later became the third Rajah, was Resident of the Third Division in 1904 and strongly disapproved of any further immigration of Chinese to the district: in his view, the Chinese were 'without being of any profit to the government [and] they were a constant source of continual trouble and annoyance to their neighbours the Dayaks and Melanaus'.[11]

The second Rajah, Charles Brooke, while not wishing to discourage Foochow immigration, accurately predicted a future of friction between the Foochows and the Ibans. In July 1908, when the Rajah appointed Baring-Gould as Resident in the Rejang, he cautioned his officer:

> There are large numbers of Chinese now living around Sibu, the Foochows and the Cantonese who are cultivating pepper and other things. They will need a good deal of looking after and encouragement—and I think some are encroaching too much on Dayak farming land which may cause trouble if not checked in time.[12]

While fully cognizant of the communal tension in the Rejang, Charles Brooke did not take any steps to resolve it. He was not prepared to risk seeing his ambitious plans of turning the Rejang into a vast agricultural Mecca ruined. In August 1908, the Rajah again wrote to Baring-Gould:

I shall be glad to hear that the question of land has been attended to as directed. If this is not properly done, I shall not be surprised that the Foochows will meet with rough treatment from either Malays or Dayaks or both.[13]

The Foochows could not understand the Iban shifting cultivation method of clearing jungle, farming it for a year or so, and then leaving it fallow and moving on to another area the next year. The Foochow method was to work the same piece of land year after year. The Iban mode of swidden farming, based on their *adat*, with obligations to spirits and ancestors, was a complex matter:

Actual family ownership is established by the felling of primary forest [*kampong*] and this is always undertaken by individual *bilek*-families. By clearing the virgin forest from a tract of land, a *bilek*-family secures full discretionary rights over the secondary jungle [*damun*] which springs up within a few months, and thereafter for as long as it remains in the community, for the initial cultivation of which it has been responsible. Once land has been acquired in this way, it becomes part of the general property of the *bilek*, and is inherited in the same way as are other valuables of the family. That is, land is inherited by the *bilek*-family as such, and all those who remain members of this local group possess parcenary rights in the land which it has acquired.[14]

In the Balleh region which Freeman was describing, where virgin land was abundant, there were no land disputes, 'little value being attached to secondary jungle; a *bilek*-family was always ready to abandon the *damun* land it had acquired, to settle in a region of untouched *kampong*'.[15] This was not the case in the Lower Rejang, where the presence of the Foochows made land a scarce and valuable commodity and there was a greater urgency for the Ibans to stake their claims. The contention over land was to get worse when the Ibans began to cultivate rubber as a cash crop. The Ibans themselves were recent immigrants to the Lower and Middle Rejang districts, having crossed the watershed from the Batang Lupar basin around the mid-nineteenth century.[16] Though they were swidden cultivators, they were more sedentary and less prone to migrating to pioneer regions than Ibans in the Upper Rejang.

In the first decade of Foochow settlement, the Ibans had not yet placed a high value on their lands, being willing to 'lend' or sell it to the Foochows: 'The Malays and Dayaks are, in most cases, willing to lend their land for one season only, but are naturally not inclined to part with it altogether, whereas the Chinese consider that one crop of padi does not compensate them for the labour they expend in ploughing and irrigation.'[17] This was the essential difference between native and Chinese cultivation. The Ibans planted for self-subsistence and any surplus was regarded as a windfall. The Foochows, on the other hand, cultivated the land with pecuniary objectives in mind, and hence treated land as an investment through which they hoped to reap good returns. Land was of spiritual and emotional value to the Ibans, while to the Foochows it was of economic importance. In 1909, when the land concession for Foochows was granted, the Foochows could formally apply for land titles. But the government had underestimated the enormous amount of work involved in surveying and laying out land boundaries and in dealing with a complex Iban *adat*. In 1910, Baring-Gould, the Resident, wrote on the progress of the Land Office:

> A number of grants under the new land regulations have been issued to the Foochow colonists. The survey work has been carried on steadily by Abang Amit, but owing to the land in many parts being indifferently cleared, progress has been slow. I propose having the land occupied by the Foochows surveyed first and got into some kind of order, as the longer it is delayed the more difficult it will be. There are numerous other applications for grants of land but they, of necessity, will have to wait as the surveyor has as much as he can cope with for some time to come.[18]

The basic problem of surveying was to remain unresolved for many years due to a lack of government staff and because of the Foochows' persistent drive to open up new land. The settlers would first clear the jungle and then apply for the land to be surveyed. It was claimed that 'in those years, one could cultivate as much land as one wanted, and then apply for it to be measured'.[19]

In 1914, some Foochow planters were reported to have encroached on Iban farming land in the Upper Binatang, without any planting permits. Two Brooke officers, H. D. Aplin and P. M. Adams, were dispatched to the Sarikei and Binatang districts to arbitrate in the disputes.[20] The land complaints placed a burden on the courts which 'have been fully occupied in land disputes and fixing boundaries; the natives and the Chinese are vying with each other to procure high land in the vicinity of Sibu which necessarily leads to disputes'.[21] High ground was favoured over low, swampy land for planting rubber trees as it was better drained. The response of the Ibans to rubber cultivation was as important a factor as their cultural attitudes towards land in their disputes with the Foochows over land. The Ibans soon realized the economic value of cultivating the land they owned rather than letting it lie fallow. Most Iban rubber smallholdings were located along the Batang Leba'an, Batang Igan, and Sungei Assan, above Sibu and in the interior, especially east of the town,[22] whereas Foochow gardens were on choice land fronting the banks of the Rejang River between Binatang and Sibu. Land competition and disputes created a heavy responsibility for the Land Office, as reflected in this 1916 report:

The Registration and Land Offices have been overtaxed. Seven hundred and thirty-five deeds were registered in the former, and large tracts of land were surveyed and grants for the same issued by the latter. . . . The present survey staff is unable to cope with the work in hand and I [Baring-Gould] very much doubt if a quarter of the land taken by the Chinese and others for planting purposes has, as yet, been surveyed. The immigration of the Chinese steadily increases and the immigrants appear to find no difficulty in obtaining work on high wages which enables them, after a few months, to start a plantation on their own.[23]

With a high percentage of Foochow-cleared land unsurveyed and unregistered, and with the Ibans challenging their rights of ownership, the Land Office was caught in a quandary. The situation was to deteriorate rapidly over the course of the next few years:

Considerable time is taken up in hearing Chinese disputes which might have been avoided had deeds been drawn up by the parties concerned, but these Chinese appear to have some objection to registering deeds, and trust to verbal arrangements which might invariably lead to the Courts and are the cause of so much ill-feeling.
 . . . Owing to the lack of a proper Land Department at Sibu, with consequently no survey, land grants have been issued promiscuously in the past. Apparently any Chinese can grab any area they please, plant it up in any way they please and then apply for a grant which is given them. The consequence is that granted land is dotted all over the country, while along the river banks, the granted land is continuous with no reserves for roads, and the owners of such land are in a position to hold all the jungle inland.[24]

As already mentioned, the Land Office, in a bid to expedite its work, and at least to register title deeds, issued 'occupation tickets' (OTs)[25] in 1920. This interim solution, with no survey work undertaken, was to make matters worse as OTs were issued too liberally, especially in the mid-1920s, and sometimes in areas just recently cultivated by the Ibans and still claimed by them. The 1923 Divisional report, in reviewing the work of the Land Office, stated:

The grants [OTs] are scattered over a large area, and no survey has been made of this district, with the result that the area and position of land available for future planting is unknown. Certain areas of lands were alloted [sic] to the Foochows for planting many years ago, but the boundaries of these are not known nor has any survey been made for them.... Arrangements have been made for a European officer to be stationed at Sibu in charge of this land office here, and this will necessitate the employment of a staff of surveyors. The lack of a survey means constant and unnecessary litigation concerning lands, and such litigation can never be finally stopped until the survey is completed.[26]

While occupation tickets constituted a major cadastral problem to be sorted out, with the Land Office not even knowing where the Foochows were planting, there was still the question, which remained unresolved, of the Foochows clearing and 'owning' land without any form of documentation. When, in 1924, the owners of rubber gardens without titles were asked to apply for occupation tickets, it was stated: 'The result was surprising, as in the last four months of the year, 1,722 gardens were reported to be containing 1,563,334 trees, of which 471,330 were being tapped; many more have still not applied for their tickets.'[27]

In the boom years between 1924 and 1925, large numbers of Foochows arrived in Sibu. They were eager to work and to start planting. The government opened up another land concession along the Binatang tributary for the Foochows. Without even surveying the area, the government proceeded to grant occupation tickets to the new arrivals. This new concession area was apparently Iban-claimed land. Expediency and revenue lay behind the government's course of action:

The Foochow Chinese are pouring into the country by every boat, and this, coupled with the high price of rubber, made it certain that if titles of some sort were not issued the land would be taken up without titles of any sort. It seemed, also, that as the newcomers were being allowed to enter the country it would have been illogical to have refused them facilities to earn their living in the only way open to them, namely, on the land, even had such refusal been practicable.

The advantages of such an issue were that the Government at least knows roughly the location of these people, in addition to which a contribution to the revenues is ensured.[28]

As expected, the Ibans protested against the Foochows moving in to clear their land. The *Sarawak Gazette*, in August 1925, reported: 'Foochows, Dayaks and Malays are still disputing over land and no satisfactory solution has as yet been found; it is necessary for an administrative and a land officer to spend some time in these places, and personally inspect the land in dispute and arbitrate as to boundaries.'[29] Even when land boundaries were staked out, the enthusiasm of the Foochow planter to occupy land beyond his boundaries could not be entirely checked. Boundaries marked with paths[30] or boundary pegs[31] could be wilfully moved.

It was not just the question of Iban customary law over and against government land grants. The rapid way in which the Foochows cleared all standing vegetation to make way for the rubber trees resulted in a process of ecological destruction and transformation which frightened the Ibans. This was how Reverend Hoover described the land rush in Binatang in 1926:

We had been given an extension on the Binatang river. This turned out to be the finest country yet opened to us. We went in with such a rush that a new situation developed in Sarawak. The old fellows with money grubstaked the new fellows on most generous terms, and they whacked in—jungle fell as though a tornado passed through. The government had experienced nothing of the kind. I was not prepared for such a development. The Dayaks were scared stiff. First they complained. Then they did everything they knew to hold the slaughter of their jungle. When the *Tuan Muda* came to Sibu, the Dayaks came in crowds and implored him to save them. The situation got so tense that the only thing left was to bring the Chinese out till boundaries could be set, and some limit put to the clearings. I am not surprised that the Dayaks were frightened—in this short time an area was cleared that looked like the beginning of an empire—500,000 para rubber trees planted, just a sample of what can be done if one has a chance.

The Rajah came to see us a few weeks ago, adjustments were made, several rivers were laid down as boundaries, and we are moving in again.[32]

The years 1925 to 1927 saw the largest number of land titles issued during the decade,[33] and a high proportion of the deeds was for opening up land in the Binatang district. Whereas in previous years, land clearings were dotted along the banks of the Rejang River, this was the first time Foochows had been so heavily concentrated in one particular area. It was an awesome sight by Sarawak standards, as plantation-sized estates, which is what these smallholdings in Binatang resembled, were not the norm. Acting District Officer Le Sueur expressed similar shock at the transformation that had taken place in Binatang, and he blamed the Foochows for their ignorance of Iban *adat*:

From the Ulu Sungei Tulai, I walked to Sungei Mador. After crossing Sungei Kara and Sungei Engkabang, near where there is a huge stretch of newly-planted Dayak padi, I reached a small belt of jungle where a number of Foochows were awaiting my arrival. As I came out of this jungle, I was astounded at the sight which met my eyes; one huge area of low hills cleared of all vegetation with occasional small patches of jungle. All this land is old farming land and the Dayaks with me pointed out several old house sites and fruit groves, many of the latter having been destroyed or partly destroyed by the Foochows' clearing and burning off.

... The whole of the trouble in this district is that they [Foochows] don't know the language or customs of the Dayaks; they come in and see old padi land which is free from all large timber and very easily cleared and naturally settle there. They do not seem to be able to distinguish between ordinary trees and fruit trees. I feel none of the damage done to Dayak fruit trees is wilful but merely out of pure ignorance.[34]

The Ibans in Binatang, unable to tolerate the situation any longer and seeing that their official protestations to the government had failed, decided to unilaterally put a halt to Foochow planting. Eight Foochows were attacked and their property was looted.[35] A number of Ibans were arrested and imprisoned.[36] The Land Office decided not to evict the Foochows, for it did not want to destroy the rubber trees. It was decreed that no more land titles were to be issued to the Foochows. Three surveyors were sent to Binatang in January 1926 to demarcate existing boundaries,[37] taking six months to complete the work.[38] The long-term effects of the Binatang incidents were more significant. Government immigration and land policies were re-appraised, and there was a conscious move to ensure that Iban 'interests' were upheld.

Government Responses: Immigration Controls and Land Laws

The immediate reaction of the Brookes to the 1925 Binatang land disputes was to introduce immigration ordinances aimed at exercising some control over the entry of Chinese into the state. In August 1925, 'a small committee of Chinese was formed after much delay, and arrangements were made to meet ships from Singapore and see that only desirable immigrants land'.[39] A month later, the District Officer reported that 'a number of Chinese who arrived from Singapore without passports were returned to that port at the expense of the *Soon Bee* shipping company'.[40] In 1926 the 'guarantee system' was applied to Foochow immigrants for the first time.[41] Under this system, no male immigrant could enter the state unless he had a guarantor. Those who failed to find work, or were found to be 'undesirables', or were discovered to be suffering from incurable diseases, were liable to be deported at the expense of the Chinese community.[42]

There is no evidence to suggest that the government strictly enforced these regulations.[43] The 1925–6 period was a good one for rubber prices. In 1925, the local rubber price averaged about $200 a *pikul*,[44] although it dropped to $120 the following year.[45] During this 'boom' period, extra Foochow immigrants were naturally appreciated, and no control over their entry appears to have been exercised. What was significant about the introduction of immigration regulations was a shift in the attitudes of the Brookes, from initially considering the Foochows as economically desirable migrants, to dealing with them far more cautiously. It was not until the Depression years of 1929 to 1932 that immigration controls were rigidly enforced.

In 1931, a major set of Land Orders was issued and, for the first time, the idea of native holdings and native communal reserves was introduced:

Any native, being a male adult, widow or spinster without adult male supporters, may hold three acres of land free of all charges [except survey fees] on which any crop may be planted, provided that no one person occupy more than one such native holding at one and the same time.[46]

In addition,

Native land reserves may be made at the discretion of the Superintendent [of Lands] by demarcation for the communal use of Malay *kampongs* and native houses, and natives may also occupy land individually by customary tenure, in both of which cases no ownership of land shall vest in the natives and in which only crops such as padi, vegetables, pineapples, sugar-cane, bananas, yams and similar cultures shall be planted.[47]

Two years later, in 1933, when the Land Settlement Order was drawn up, more importance was attached to native interests. Native customary rights were recognized with respect to 'land planted with fruit trees, land in continuous occupation or which has been cultivated or built on within three years'.[48] If a native was able to establish his customary rights over any land, the Settlement Officer of the Land Office could either issue him with a title or pay him compensation money. To prevent the 'lending' and selling of native land to Foochows, native holdings 'would be indivisible and would not be charged, sub-leased, transferred, or otherwise disposed of except by inheritance'.[49]

The 1931 and 1933 land laws, drawn up with the idea of separate ethnic residential areas, provided the basis for the 1948 Land Ordinance which classified all land in Sarawak into the following groups: (a) mixed zone land, (b) native area land, (c) native customary land,

(d) reserved land, and (e) interior area land.⁵⁰ The effect of the 1948 legislation was to restrict non-natives to 'mixed zone land' areas. 'Native' was not explicitly defined, but was taken to exclude Chinese according to the Interpretation Ordinance of 1933.⁵¹ Mixed zone areas were generally in towns, in peripheral urban locations, and in rural areas where there were concentrations of Chinese. Since 1931, the Chinese have not been able to cultivate land of their choice unless authorized to do so.

Economic Background

Boom and Slump

Occurring contemporaneously with political changes in the attitudes of the Brookes towards the Foochows were economic changes resulting from the cyclical rhythm of prosperity and hard times which affected the Foochows.

When rubber prices fell sharply in 1920 and 1921, many Foochows and Henghuas were in trouble. The government offered to alleviate the situation by creating work for the unemployed:

> Towards the end of the year [1920], the government employed hundreds of Chinese [Foochows and Henghuas] who were out of work. For a time, the average number working was about 300 a day. New roads were constructed from Lanang Road to Bukit Lima, from Sungei Merah to the Igan, from the Fort to No. 1 Bungalow, and from the Fort to Central Road.⁵²

Many families were unable to obtain food, and a lot of *sinkheh* who were hired as labourers in rubber gardens were forced to leave their jobs. The general state of despair led to 'a slight disturbance in the bazaar, with some minor looting'.⁵³ The headmen of the individual Foochow districts were asked to send in lists with names of those in need of assistance. Each applicant was examined as to his financial position, with preference given to those with wives and families. Several hundred men were employed to work on the roads. Fifty needy bachelors were sent to the Sadong colliery in Simunjan.⁵⁴ In the meantime, there was an outflow of Foochows from the Rejang. It was said that 'by every trip of the Flevo, large numbers are returning to China'.⁵⁵ The flow of Foochows out of Sibu was influenced by rubber prices.

By 1924, the number of new arrivals had begun to pick up again, with 1,412 recorded entries as compared to 491 the previous year. From 1924 to 1929, the number of new settlers rose steadily. Between these years, there was some fluctuation in average rubber prices, but

TABLE 8.1
Foochow Arrivals, Land Titles, and Rubber Prices, 1921–1940

Year	Rubber Prices (Per Pound) s.	d.	Foochow Arrivals Men	Women	Children	Total	Land Titles
1921		9.50	n.a.	n.a.	n.a.	114	47
1922		9.30	35	13	16	64	38
1923	1	3.30	377	100	14	491	19
1924	1	7.80	1,112	282	18	1,412	18
1925	2	11.00	1,939	251		2,190	892
1926	1	11.75	2,131*	477*	777	3,385*	892
1927	1	6.80	1,882*	433*	538	2,853*	1,010
1928		10.75	1,207*	507*	573	2,287*	401
1929		10.25	1,511*	511*	372*	2,394*	483
1930		5.90	n.a.	n.a.	n.a.	n.a.	566
1931		3.50	n.a.	n.a.	n.a.	n.a.	161
1932		2.30	n.a.	n.a.	n.a.	n.a.	117
1933		3.25	n.a.	n.a.	n.a.	n.a.	256
1934		6.25	n.a.	n.a.	n.a.	n.a.	105
1935		6.00	n.a.	n.a.	n.a.	n.a.	52
1936		7.75	n.a.	n.a.	n.a.	n.a.	31
1937		9.50	n.a.	n.a.	n.a.	n.a.	91
1938		7.20	n.a.	n.a.	n.a.	n.a.	66
1939		9.00	n.a.	n.a.	n.a.	n.a.	47
1940	1	0	n.a.	n.a.	n.a.	n.a.	111

Sources: SG, 1921–9; Andrew McFadyean, *The History of Rubber Regulation 1934–1943* (London: George Allen and Unwin, 1944), p. 239; and Land Office files, Sibu.
*Includes Hokkiens, Henghuas, Hakkas, and Cantonese.

it did not appear to have much impact on the inflow of Foochows. This could have been because rubber cultivation was a long-term investment and its effects would only be felt after a period of time. It was when rubber trees matured and prices fell that the planters would feel the impact and become discouraged, prompting some to sell their property and leave the country. On the whole, as more Foochows entered the country, more land titles were taken out, as is evident from Table 8.1.

The increasing tempo of Foochow settlement affected the urban economy of Sibu town and caused some structural economic changes among the Foochows. A *Sarawak Gazette* report in October 1924 commented on the changes in Sibu:

... Sibu itself is expanding quickly; bazaar shops are going up every month; roads, drains, water and lighting are developing rapidly under the lead of the Resident, supported by the principal Chinese of the place. The business activity of the Sibu Chinese has stimulated trade with the Dayaks, which coupled with the activities of the Foochows, has increased the exports and imports of the Division enormously in the last few years.[56]

Trade in Sibu town expanded with the Foochows supplying rubber to the shops and the Ibans delivering jungle products.

In 1925, rubber prices averaged $200 per *pikul*, and an acre of rubber garden was estimated to be worth between $500 and $600. Foochows competed with each other for rubber gardens that were up for sale, or else applied to the Land Office for land titles. Some enterprising Foochows diversified their interests or switched occupations from farming to trading. Companies were started to deal in groceries, piece-goods, household effects, medicine, and gold and silver jewellery, and some Foochows put their capital into sawmills, ricemills, rubber processing plants, and shipping companies.[57] The shops in Sibu, traditionally run by Hokkiens and Teochius, were soon predominantly Foochow-owned; by 1928, there were only 40 to 50 shops owned by Hokkiens and Teochius, while Foochows owned about 160 to 170 shops.[58] Despite the start of occupational diversification, the majority of Foochows remained rubber smallholders. With the onset of the Depression in 1929, rubber prices dropped from $48.85 per *pikul* in 1929, to $25.72 in 1930 and $13.30 in 1931, right down to $6.95 in 1932,[59] and Sibu's trade in rubber collapsed.

The fall-off in rubber prices once again caused a great deal of unemployment among the Foochows. A large proportion of the unemployed Foochows were smallholders, whose rubber gardens were not mature yet and who relied on part-time employment from other owners. Unemployed Foochows were set to work on the public roads. This was only a temporary measure. In 1931, as the employment situation worsened and as it was no longer economically expedient to tap rubber trees, the government began the drastic measure of 'repatriating all those Chinese whose age and physical condition made it improbable that they could find work'.[60] In 1932, 358 Chinese were repatriated.[61] There is no breakdown of figures for those sent by Division or town, but the majority were from Kuching and Sibu, the two urban centres with substantial concentrations of jobless Chinese. It was a programme which had the active support of the Kuching Chinese leaders. The Kuching Chinese Unemployment Relief Fund Committee paid half of the expenses of repatriation, and at the same time it was responsible for screening applicants for selection.[62] This measure, together with the policy of restricting immigration, was successful in keeping the number of Chinese immigrants down for a period of five years, from 1929 to 1933.[63] In the mid-1920s, when rubber prices had not yet dropped drastically, immigration controls were not rigidly enforced.

Rubber prices picked up again at the end of 1933, doubling from $12.66 a *pikul* in 1933 to $23.73 in 1934.[64] Foochows and Ibans responded to the new prices by 'abandoning the padi-fields which they had to some extent relied on for a living during the past 3 years'.[65] This shift in attention from rice to rubber was widespread throughout the state, evident in the figures for rice imports, which almost doubled in two years, from 279,928 *pikul* in 1933 to 551,324 *pikul* in 1935.[66] In anticipation of better prices, 'a large number of owners of land already planted with rubber applied for land titles'.[67] For the first time since the Depression, there was a demand for labour in the rubber gardens.[68] However, this did not change the Brooke attitude towards Chinese immigration. Immigration laws continued to be enforced and were supplemented by the enactment of Order No. D-2 (Disembarkation), according to which aliens would be required to pay a deposit before being permitted to land. An immigrant had to be in possession of a Sarawak certificate of identity, passport, or landing permit.[69]

The reason that immigration controls were not relaxed was because Sarawak, in 1934, became a signatory to the international rubber restriction scheme, and any removal of immigration limits would have defeated the objective of reducing rubber output.

Rubber Restriction

When rubber restriction was begun in 1934, there was a prohibition, of four months' duration on the collection, storage, and export of scrap rubber.[70] In 1935, during the months of July, October, and November, there was a complete tapping holiday of 44 days. Bridges, the official responsible for the enforcement of restriction measures, reported that 'tapping holidays were faithfully observed throughout the country which curtailed production by approximately 1,800 tons'.[71] A deterrent to tapping was the informant network which the Brookes encouraged, giving rewards to persons who were willing to report on violations.[72] More hardship was caused to rubber tappers, when, in 1936, five tapping holidays of 114 days' duration were instituted.

Enforced rubber output reduction was used in conjunction with immigration restraints, so that the combined effect of low rubber prices and immigration controls resulted in more Chinese leaving than entering the state between the years 1930 and 1937.[73] In fact, immigration control was seen as a necessary adjunct towards lowering rubber production in the state. The fall-off in the number of Chinese allowed into Sarawak appeared to have caused a labour shortage in the rubber gardens by 1936. It was stated: 'In some areas, Chinese tappers on a share basis were asking and getting 60 per cent of the crop, or wages of up to $2 per day.'[74] Bridges described immigration cut-backs as the worst feature of the rubber restriction scheme.[75] In spite of the manpower shortage, there was no indication that immigration restrictions would be lifted. The 1937 Administration Report affirmed: '... the policy of rigidly restricting male immigration was continued for the greater part of the year, since it was considered that control should not be relaxed until it could be ascertained whether the introduction of the coupon system would result in the release or absorption of labour.'[76]

The coupon system was used because Bridges was sceptical of the value of tapping holidays. Officials from Malaya helped to count and survey more than 90,000 smallholdings in the country, and each owner received coupons entitling him to market a specified amount of rubber in accordance with the size of his garden.[77] In addition, under the terms of the 1934 international agreement, all new planting was prohibited.

Rubber restriction and immigration restraints were two forms of government pressure which helped to instil in the Foochows an increased sense of community and an awareness of the need to protect the interests of its members. The Foochows faced isolation and rivalry from other Chinese, the Hokkiens and the Cantonese, and they also encountered increasing tension in their relations with the native Ibans. Community self-help in this difficult period is best illustrated by the strength and work of the Methodist Church and in the growth and proliferation of community associations in the 1930s.

Social Organization

The Methodist Church

Just after 1920, there was a rapid expansion of the Methodist Church[78] due to a combination of windfall prosperity and a growth in population.

Between 1921 and 1930, fifteen churches were built. In 1928, Hoover had a retinue of twenty-three lay preachers: 'To all, I give a small subscription, usually ten to twenty dollars [Mexican] monthly, depending on the circumstances. This money comes mostly from rubber gardens owned by the Mission.'[79]

By 1928, there were twenty-five chapels. Each of these stations was supported by its own rubber garden.[80] This was how Hoover described the expanding work of the Methodist Mission in 1934:

New places are popping up here, there and everywhere, some with names, and some without—just clearings in the jungle. The people want schools and churches, give them the land, build the house and pay the preacher–teacher, asking the Mission to help with only $100 with the building, and $5 a month on salary. In a short time, there is a prosperous ever-growing community and we are the centre of it. All that the Mission now needs is $200 a month to support the entire work.[81]

The Foochows themselves played an active role in the church and church work, unlike their Anglican and Catholic brethren elsewhere in the state who relied heavily upon foreign missionaries and workers. The Methodist Church, for example, had a growing band of local preachers, variously estimated in the mid-1930s at 113[82] and 125.[83] The lay preacher's position in the Foochow community was a socially prestigious one, and this may explain the popularity of preaching. Church-going on Sundays was strictly adhered to,[84] and is observed to this very day. Hoover single-handedly had the task of co-ordinating church work and of giving encouragement to each of the missions. As most of the stations were in remote and inaccessible locations, he could not visit his stations more than twice a year. His arrival was always eagerly awaited and treated as a momentous occasion:

... All babies are brought forward for baptism, new members are taken into the church and the holy communion is administered. It takes nearly four months to cover the circuit, part of it by steam boat, part of it by motor launch, part by bicycle, and much of it on foot through long stretches of plantation or jungle.[85]

The missionary did his best to encourage the Foochows to take an interest in their religion, although his attitude to the material benefits of Christianity might even in the 1980s be regarded as somewhat unorthodox. Hoover stated quite explicitly in his biography that he hoped the Foochows would see the economic advantages associated with becoming members of the Church:

I don't want them to become Christians for pay, but I do want them to see that, other things being equal, it pays to be a Christian. There's no sin in having money. The sin comes when one loves money, worships it as an idol in a temple. If I can help these half-starved folks to reach a higher economic plane, I am doing some good, and if I can do the much better work of teaching them on that new level, to share their money with the needier, I am doing a lot of good.[86]

Not without some justification, Hoover was to boast, 'our missionary business included the whole range of business interests, religion, education, politics, medicine, immigration, town planning, road building, machinery, boats, etc.'[87] However, with the growing prosperity of the Foochows, there were signs of decreasing dependence on the Church as the pillar of their social organization. Other forms of social organization had begun to supplement the Church.

Associations

The more complex nature of Foochow social organization was to be found in associations which catered for particular needs and for groupings of people of similar county origin. The first pan-Foochow, non-lineage dialect association was formed in 1902; it was called the Kwang Yung Serh Association (Bright and Far Association). Indeed, this was the only Foochow association until the 1920s, probably because the Methodist Church already provided an active community social life and a welfare support system for the Foochows. Another reason could be that, in the formative years of the colony, the Foochows had not yet experienced the effects of external pressures such as land disputes, the fluctuation in rubber prices, and land and immigration controls, as they were to do in the 1920s and 1930s, and therefore had no urgent reason to form associations representing their particular interests. A final factor could be that the dispersed settlement pattern of the Foochows in rural locations did not make it easy for them to get together to set up formal associations.

However, by the early 1930s, with the diversification of the Foochows into urban occupations and with an expanding population which had changing educational and social aspirations, more associations began to be formed. In 1926, the Kwang Yung Serh Association was formally registered as a benevolent body whose main objectives were helping the infirm and needy, repatriating the elderly, and buying coffins for the destitute who had died. A pan-dialect group association, the Sibu Chinese Chambers of Commerce, which included Foochows, was formed in 1930.[88] The Kutien Association for

Foochows of Kutien county was formed a year later.[89] Various other associations which took in Foochows were the Sibu Chinese Club and the Overseas Chinese Sporting Club, both established in 1932 for social and recreational purposes. The Kwang Hua School Alumni Association at Sadit was set up in 1934, and the Kwang Nam School Alumni Association in 1936. To cater specifically for the Foochows, the Sibu Foochow Commercial Recreation Club was established in 1936 with the objective of promoting and improving trade for its Foochow members, in contrast to the Chinese Chamber of Commerce which served the interest of all dialect groups.

Not to be left out, the Methodist Church decided to undertake welfare work by forming the Sibu Methodist Philanthropic Association for the purpose of 'burial affairs' in 1933, perhaps to compete with the Kwang Yung Serh Association. Another benevolent association was the Lok Sien Association in Sadit, registered in 1935. 'Old boy' networks seemed to be a favoured way of formalizing social ties. The Tung Hwa Alumni Association was set up in 1939. Associations of this type perhaps represented members' pride in their educational background, but it also meant that educated Foochows were prepared to represent and uphold wider community interests. The Mintsing Association for Foochows hailing from Mintsing county was formed just before the Second World War.[90]

By the end of the Brooke dynasty, the Foochows had become a well-established community in the Rejang. They had overcome pioneering set-backs, and had grown into a prosperous community as a result of rubber and Methodism, but at the cost of deteriorating ethnic relations.

1. In March 1901, two Cantonese individuals, Tang Kung Shiok and Chiang Cho Shiang, signed an agreement with Charles Brooke to bring in a group of Cantonese settlers to reside at Sungei Lanang, a few miles up-river from Sibu. The Cantonese and the Foochows did not get along with each other and clashed right from the start. Information from Kang Chung Siew, 19 May 1981, Sibu; *SG*, 2 June 1902, p. 125; *SG*, 2 June 1903, p. 124; *SG*, 1 June 1905, pp. 135–7; *SG*, 1906, pp. 140–1; and *SG*, 17 August 1908, pp. 208–9.
2. Personal interview with Sim Ngiik King, 23 June 1981, Binatang.
3. *SG*, 2 January 1915, pp. 6–7.
4. Lau Tzy Cheng, *Wong Nai Siong and the New Foochow* (Singapore: Nanyang Institute, 1979), p. 52 (text in Chinese).
5. Interview with Wong Ngiong Hua, 25 June 1981, Sarikei.
6. Information from Hii Hua Siong, 27 June 1981, Sarikei.
7. Lau, *Wong Nai Siong and the New Foochow*, pp. 16–18.

8. Diu Mee Kuok, 'The Diffusion of Foochow Settlement in the Sibu–Binatang Area, Central Sarawak, 1901–1970' (MA thesis, University of Hawaii, 1972), p. 60.
9. *SG*, 2 September 1901, p. 102.
10. Frank T. Cartwright, *Tuan Hoover of Borneo* (New York: The Abingdon Press, 1938), p. 112.
11. *SG*, 4 June 1904, pp. 124–5.
12. Charles Brooke, 24 July 1908, Letters, Vol. 6, p. 81, SA.
13. Charles Brooke, 2 August 1908, Letters, Vol. 6, p. 84, SA.
14. Derek Freeman, *Report on the Iban of Sarawak* (Kuching: Government Printing Office, 1955), p. 143.
15. Ibid.
16. See Robert Pringle, *Rajahs and Rebels* (London: Macmillan, 1970), Chapter 8.
17. *SG*, 16 June 1909, p. 17.
18. *SG*, 5 January 1910, p. 10.
19. Information from Ling Chu King, 21 May 1981, Sibu.
20. *SG*, 2 January 1915, pp. 6–7.
21. *SG*, 17 March 1916, p. 52.
22. Vinson Sutlive, 'From Longhouse to Pasar: Urbanization in Sarawak, East Malaysia' (Ph.D. thesis, University of Pittsburgh, 1972), p. 32.
23. *SG*, 1 March 1917, p. 59.
24. *SG*, 1 November 1920, p. 241.
25. Once land titles were issued, the Iban concept of customary law was thrown into confusion. Court decisions tended to favour legalized land procedures as this example illustrates:
    In September 1920, three Foochows, Tiong Khoon Ping, Wong Ping Chuan, and Wong Ah Hur, accused Sunggang of removing boundary pegs from their land to which they had been granted titles. *Abang* Abdul Hamid, Chief Surveyor, examined the land in dispute:
'It is quite clear that the defendant has taken land included in grants No. 579 and 519. Defendant has no claim at all on the land. Tiong Khoon Ping says that the defendant has planted 2 *mulongs* [sago palm] on his land, these of course the defendant will lose. The defendant is warned not to trespass on granted land. Defendant is to pay all costs of the case and for the new *belian* marks to be put in. Defendant also loses any work he may have done on the granted land, and also anything he may have planted.'
'Tiong Khoon Ping, Wong Ping Chuan and Wong Ah Hur v. Sunggang of Kampong Nangka', 9 September 1920, Sibu CCB, p. 291, SA.
26. *SG*, 1 May 1923, p. 136.
27. *SG*, 1 May 1924, p. 157.
28. *SGG*, 16 August 1926, p. 201.
29. *SG*, 3 August 1925, p. 196.
30. 'Serin of Telok Selalo v. Wong Eng Guan (Foochow), Pulo Kerto', 4 March 1919, Sibu CCB, p. 62, SA.
31. Kirak v. Ah Chai, Kanowit, 10 March 1919, Sibu CCB, p. 71, SA.
32. Singapore Methodist Conference, Minutes of Thirty-fourth Session, 1926.
33. In 1925, 892 land titles were issued. The same number of land grants was issued in 1926. The number rose to 1,010 titles in 1927. *Source*: Land Office files, Sibu.
34. *SG*, 1 December 1925, pp. 316–17.
35. *SG*, 1 October 1925, p. 263.
36. *SG*, 1 February 1926, p. 38.
37. *SG*, 1 April 1926, p. 98.

38. *SG*, 1 October 1926, p. 268.
39. *SG*, 1 October 1925, p. 263.
40. *SG*, 1 December 1926, p. 324.
41. *Sarawak Administration Report for 1930*, p. 52.
42. *SG*, 2 May 1927, pp. 122-3.
43. According to information obtained from Wong Kwong Yu, 30 May 1981, Sibu.
44. Quoted by Kho Cheng Kang, 19 May 1981, Sibu.
45. *SG*, 1 April 1926, p. 89.
46. T. Stirling Boyd (comp.), *The Laws of Sarawak, 1927-1935* (London: Bradbury, Wilkinson, 1936), p. 251.
47. Ibid., p. 252.
48. Ibid., p. 300.
49. Ibid., pp. 300-1.
50. A. F. Porter, *Land Administration in Sarawak* (Kuching: Government Printing Office, 1967), p. 60.
51. Interpretation Order No. 1-1, 1933, in Boyd, *The Laws of Sarawak, 1927-1935*, p. 207.
52. *SG*, 1 April 1921, p. 89.
53. Ibid.
54. Ibid.
55. *SG*, 1 February 1921, p. 13.
56. *SG*, 1 October 1924, p. 310.
57. Chiang Liu, 'Chinese Pioneers, A.D. 1900: The New Foochow Settlement of Sarawak', *SMJ*, Vol. 6, No. 6 (December 1955), p. 541.
58. Ibid., p. 542; and *SG*, 1 May 1928, p. 93.
59. *Sarawak Administration Report for 1929*, p. 3; *Sarawak Administration Report for 1930*, p. 4; *Sarawak Administration Report for 1931*, p. 3; and *Sarawak Administration Report for 1932*, p. 3.
60. *Sarawak Administration Report for 1931*, p. 26.
61. *Sarawak Administration Report for 1932*, p. 14.
62. *Sarawak Administration Report for 1931*, p. 26.
63. *Sarawak Administration Report for 1933*, p. 14.
64. *Sarawak Administration Report for 1933*, p. 3; and *Sarawak Administration Report for 1934*, p. 3.
65. *Sarawak Administration Report for 1935*, p. 41.
66. Ibid.
67. *Sarawak Administration Report for 1934*, p. 4.
68. Ibid., p. 16.
69. Ibid., p. 17.
70. Rubber Restriction Order No. R-3, 1934, in Boyd, *The Laws of Sarawak, 1927-1935*, p. 504.
71. W. F. N. Bridges, 'A Report on Rubber Regulation in Sarawak' (Kuala Lumpur: Library of Rubber Research Institute, Kuala Lumpur, 1937), p. 16 (typescript).
72. *SG*, 2 May 1938, p. 64.
73. See *Sarawak Administration Reports*, 1931-8.
74. Bridges, op. cit., p. 19.
75. Ibid.
76. *Sarawak Administration Report for 1937*, p. 14.
77. *SG*, 2 May 1938, p. 102.

78. In 1925 and 1930, the Methodist Mission acquired 15 acres and 25 acres of land respectively, presumably for rubber planting. *Source*: Land Office files, Sibu.

79. Cartwright, op. cit., p. 162.

80. Ibid., p. 163.

81. Nathalie Means, *Malaysian Mosaic: A Story of Fifty Years of Methodism* (Singapore: Methodist Book Room, 1935), pp. 125–6.

82. *BB*, 5 September 1981, p. 22.

83. Means, op. cit., pp. 125–6.

84. Information from Ting Ming Hoe, 19 May 1981, Sibu.

85. Means, op. cit., pp. 125–6.

86. Cartwright, op. cit., pp. 169–70.

87. Ibid., pp. 122 and 166; *SG*, 17 August 1908, pp. 208–9; and 2 June 1913, p. 118.

88. *SGG*, 3 January 1933, p. 8.

89. Information on this and other associations is obtained from *SGG*, 3 January 1933, pp. 7–8; 16 February 1938, pp. 71–3; and 1 March 1940, p. 211.

90. Lau, *Wong Nai Siong and the New Foochow*, p. 42.

9
Chinese Labourers and the State Economy, 1870–1941

THE conventional historiographical view has been that the Brookes did not actively encourage commercial enterprise in Sarawak[1] for the altruistic reason of protecting native interests, that they wanted to preserve the Raj the way they found it,[2] and that they were jealous of and feared other European interests. This view of the Brookes' attitude to commercial development is hardly accurate. An account has already been given in an earlier chapter of how the White Rajahs encouraged the Borneo Company in the later half of the nineteenth century at the expense of Hakka mining interests in Bau. The Brookes did indeed play a major commercial role by opening up coal mines at Sadong (Simunjan) in the 1870s and at Muara (Brooketon) in 1884, as well as running several experimental tobacco, tea, and coffee estates.[3] In 1909, the Anglo-Saxon Petroleum Company (later known as Sarawak Oilfields Limited and Shell Company) was given a free hand to extract oil deposits. A large and steady work-force was needed to labour on the coal mines and oilfields. Local natives were not considered suitable for these tasks. Just as the tin mines and rubber estates of Malaya turned to Chinese and Indian labourers, so too did Sarawak recruit Chinese *sinkheh* labour through Singapore or directly from China. Coal was a leading mineral export from 1889 to 1898, and oil from 1920 till the outbreak of the Second World War. It was in large measure the unrecognized efforts of Chinese labourers which made coal and oil such important fuels for local use and export.

The Opening of the Sadong Coal Mines

The Sadong colliery was situated some 3 miles east of the present Simunjan bazaar at the western end of Mount Ngili. Coal deposits were known for a long time to have existed there, but it was not until 1857 that mining was first begun. The Borneo Company's venture in digging the ore failed: 'This attempt to open a coal mine in Borneo was a costly failure ... rumours said 20,000 pounds ... all because to save a few hundred pounds, the opinion had been taken of a practical miner ... instead of obtaining a report from a scientific engineer who could have discovered if a workable seam existed.'⁴

The coal seams were seemingly unworkable and the idea of mining the coal was abandoned.⁵ It was left to Charles Brooke to reopen the mines with the help of Chinese labour in the early 1870s.

In 1872, a Chinese party was engaged to work the coal and deliver it to the Brooke government in Simunjan for $4.00 a ton. The *Sarawak Gazette* of 31 January 1872 stated: 'There is reported to be a large quantity which may be had by opening up the seam round the hill and it is expected that the supply will last for a considerable time, but of course without any European to supervise, it will not be worked on a scientific principle.'⁶

The implication of this somewhat patronizing *Sarawak Gazette* statement was that the Chinese had not mined successfully elsewhere, when European supervision and technology were absent. In the hope of attracting European speculators, an advertisement appeared in September 1872 in the *Sarawak Gazette*: 'The coal seams of Simunjan in the Sadong and Silantek in the Batang Lupar, varying in thickness between 4 and 5 feet of exceedingly good quality of gentle angle with firm roofing [so far as can be ascertained] may be worked by an individual or company on a payment of 10 cents a ton brought to grass.'⁷

No company responded to this advertisement and the Brookes then reluctantly contracted in 1873 with a Chinese party to work the coal seams. At this stage, there was only one wooden tramway leading from the side of Mount Ngili. A substantial amount of coal must have been mined under this contract, for it was reported that there was a surplus for export after the government steamers had been supplied.⁸ The area worked expanded considerably, and shafts were driven underground into the face of the mountain. By 1876, the Brookes' interest in coal had increased to the extent that the Rajah invested £4,000 in the construction of a railway and the purchase of a locomotive, doing away with the old method of using a wooden tramway whose wagons were pushed and drawn by Chinese labourers and buffaloes. It was optimistically expected that 300 to 600 tons of coal would be exported a month.⁹ With such heavy capital expenditure and high expectations of a profitable return on investment, the next problem was to procure the skilled labour to do the difficult and dangerous work below ground.

The Procurement of Chinese Labourers

There were initial difficulties in obtaining coolies from either Singapore or China, as it was widely believed that indentured labourers were sold off as slaves in Sarawak. Coolies from China were paid advances of between $25 and $30, and wages of $8 to $10 per month.[10] In 1886, out of a labour force of 120, only half were Chinese, the rest being Malays and Ibans.[11] Charles Brooke responded to the slavery charges by inserting this notice in the *Sarawak Gazette* in August 1882:

Whereas it has been reported to the Sarawak Government that reports are spread in Singapore by mischievous people to the effect that Chinese coolies on arrival in Sarawak are sold off into slavery, notice is hereby given that slavery amongst the Chinese is forbidden by law in Sarawak territory and that Chinese coolies on arrival obtain work either on gambier and pepper gardens, roads, public works and etc.[12]

Sarawak recruited labourers for the Sadong coal mines through the official agency of the Chinese Protectorate in Singapore. The White Rajah had his own counter-charges to make against the accusations of slavery by claiming that unhealthy and unfit men were sent by the Protectorate:

... Men of any description are taken, sound or unsound, diseased or otherwise, and delivered to consignee in exchange for dollars. The consignee on receiving his men finds many of them useless and quite unfitted [sic] for work, but on remonstrance is informed that he must take what is sent him, serviceable or otherwise, and is made to feel the ring of coolie recruiters in Singapore is too strong for him, and is backed by the red tapeism of the Chinese Protectorate there.

Out of some 300 coolies recently obtained by the Sarawak Government to work in the Sadong coal mines, one man died on the voyage over. Of the rest who were transhipped from steamer to steamer for delivery at their destination, six died or almost immediately upon landing, mostly from chronic dysentry [sic], and some scores were so weak or debilitated or in so advanced a stage of disease as to have been incapable of doing a single day's work since arrival. Two also of the gang were absolute lunatics and one stone blind.

Of coolies in a diseased condition who were considered hopeless cases by the Principal Medical Officer here, twelve who did not die were sent to Singapore with a request for compensation from the recruiters, and these on being taken before the Protectorate one and all attributed their condition to the influence of coal amongst which they said they worked.[13]

Charles Brooke shifted the blame for the wretched condition of the *sinkheh* away from the hazards of working in the mines to the inept screening and recruiting procedures of the Chinese Protectorate in Singapore. He refused to acknowledge that poor working conditions at the Sadong coal mines were responsible for the illnesses and deaths of the men. The Protectorate accepted the Rajah's claims but threatened to boycott Sarawak. A result of the exchange of charges and counter-charges made by the Rajah and the Protectorate against each other was an improvement in medical facilities on both sides. The Singapore Medical Officer began examining out-going coolies in the 1890s. The Rajah constructed coolie depots in Kuching, and in-coming vessels were instructed not to discharge their passengers until they had been medically examined.[14]

Every effort was made to increase the number of Chinese coolies on the mines. The Resident in Kuching, on his trips to the mines, would sometimes personally bring coolies with him. In February 1888, twenty Hainanese labourers accompanied the Resident to the mine.[15] Plans were broached in June 1888 to run a direct steamer service between Simunjan and Singapore to carry coal over to Singapore, and it was recognized that more coolies would be needed to ensure a steady supply of coal.[16] By 1891, the labour force in the Simunjan colliery numbered 800 men,[17] but the turnover was high. Dissatisfied with the services of the Protectorate, the Rajah appointed agents to handle the recruitment of labourers. A. L. Johnston and Company acted as the handling agent until October 1889, when Paterson Simons and Company took over.[18] Despite the constant search for workers, the colliery management was to admit in 1896 that 'labour had been most difficult to obtain'.[19] The reasons for this were the poor working conditions on the mines.

Working and Living Conditions

The coal was dug from the sides of Mount Ngili, and from underground by a network of tunnels. By the end of the nineteenth century, after the richest and easiest seams had been worked, only thin coal seams remained and working conditions became more dangerous, with the very real possibility of shafts caving in and roofs collapsing:

> The present strata where coal is now being worked, is in a very uncertain and disturbed condition, the coal is very thin and constantly intersected by dirt bands; the dip of the strata is very variable, the average being 4 inches per yard, but at times, this will alter to as much as 8 or 9 inches, the roof above the coal is in a very broken and tumbled condition giving endless trouble to support.[20]

In the 1920s, two alternate day and night shifts of eight hours' duration each was in force.[21] The coal was transported from the hill to the waterfront by locomotive. From the waterfront, the coal was loaded on to waiting ships. A single locomotive was unable to cope with the transportation of the ore.

The labourers were housed in 'coolie lines', longhouse-style barracks. They could either be paid monthly wages—which, in 1894, came to $8–10 per month, with food, clothing, and accommodation provided—or else they received payment according to output on a piece-rate basis.[22] At least until the end of the nineteenth century, most of the coolies were paid wages.

A good indication of working and living conditions for the coolies comes from the sporadic medical reports in the *Sarawak Gazette*. The labourers worked long hours in poorly lit and badly ventilated tunnels. Living conditions in the 'coolie lines' were abysmal, with the houses subject to flooding. It was only in 1906 that the coolie lines were demolished and shifted to higher ground. The food was inadequate and hardly nutritious, and outbreaks of beriberi due to dietary deficiencies were common.[23]

In the first quarter of 1889, out of thirty-eight miners who were treated for various ailments, thirty were found to be suffering from ulceration[24] of the lungs caused by insufficient ventilation and damp conditions in the tunnels. Poor sanitation and unhygienically prepared food were responsible for a scourge called *dochmius duodenalis*, a hookworm disease. Thirteen cases of the disease were discovered in the first quarter of 1891.[25] This ailment persisted until the 1920s. It was only as late as 1923 that the first steps were taken to remedy this malady. A medical officer, Dr O'Driscall, spent twenty-three days in the Sadong investigating the causes of *dochmius duodenalis*. Following his investigations, more latrines were built, incinerators were provided, and sanitary *mandor* and coolies were appointed.[26]

The medical reports in the *Sarawak Gazette* were unusually frank on the physical condition of the Sadong colliery workers. The second quarterly report for 1891 stated:

Sadong coolies continue to arrive from the mines in a terribly broken-down state of health. Many, according to their own accounts, have been unable to do any work for 2 to 3 months before being sent to Kuching for treatment; one died in the hospital on the day he arrived in Kuching ... 21 Sadong coolies were treated and paid $79.10 out of the total fees of $306.88.[27]

Many patients were sent to Kuching to be treated. Out of 336 cases treated at the Kuching General Hospital in 1891, 54 were Sadong mine labourers.[28] This did not include the workers who were treated locally at the colliery. A hospital was constructed at the colliery in 1894[29] to cope with the casualties and injuries which had arisen during the peak years of coal production. Over the next two years, 1895 and 1896, Sadong, among all the outstations under the Brooke administration, received the maximum allocations in medicine and drugs, amounting to $146.40 and $123.90 respectively.[30] This was due to the high accident rate and the high incidence of disease and sickness at the mines. Three hundred and eighty coolies were treated at the hospital in Sadong during the year 1898.[31] Overcrowding at the hospital was a problem. During a visit by the Principal Medical Officer, A. J. G. Barker, to the hospital in August 1898, thirty sick patients were occupying beds meant for ten persons.[32]

A serious occupational hazard came in the form of accidents. On 6 April 1889, three workmen on a night shift suffocated to death after inhaling poisonous gas:

> It seems whilst at night work in No. 3 mine, they struck an abandoned and apparently unknown alley way, and the water commencing to flow in upon them, the rest of the miners fled to the surface being frightened by the sound of water, and they called out to the three remaining men to follow ... they thought the inrush of water to be insignificant and preferred to remain and finish their night's allotment, and accordingly refused to leave it, and upon the poisonous air reaching them they were apparently speedily overcome and suffocated to death.[33]

The mine management did not take any safety precautions of any kind, nor was any compensation given for deaths or accidents. Instead, the mine employers blamed the labourers for ignoring their own safety when they had to work in dangerous circumstances. If the coolies refused to work under trying conditions, they were accused of being lazy or insubordinate. No proper inquests were conducted when miners died. In 1922, a labourer, one Tsang Ngee, was found dead in a disused part of the mine, and the manager conveniently recorded a verdict of death by misadventure due to 'deceased being subject to fits'.[34]

A letter of complaint written by a newly appointed mine manager, M. Kendall, in 1904, to the Resident in Kuching reveals much about the dreadful physical conditions of the tunnels and the underground work:

> I am writing to you in regard of No. 4 and No. 5 mines, Sadong, in which I am in charge of, stating the bad condition in which I found both mines, with the intake and return airways as bad as they could be. Both mines are full of black damp, and scarcely a lamp would burn, and here the water in these mines is up to my waist which I have to travel through everyday; between the black damp and the water is enough to kill any man.[35]

For the Chinese *sinkheh* who were recruited to work in the mines, few options were open. To protest against the appalling working conditions, many left their work altogether (in official language, 'absconding'), struck back at tyrannical managers, or went on strike. If they continued working, there was little hope of improvement in their material circumstances, and if they were unable to cope with the job, they were discharged. For the desperate, one means of escape was death, or—less drastic—illness or injury. One of the first professional lawyers to go to Sarawak in the early 1930s, K. H. Digby, described the system of procuring indentured labourers as little better off than slavery, indentured labourers being entirely under the direction and control of their 'owner'.[36]

Many coolies simply absconded to escape from an oppressive working environment. 'Absconding under advances' was treated as a criminal offence carrying severe penalties 'although no support for this view could be found in the law'.[37] The punishment would be imprisonment and rattan flogging, and if there was an immediate need for the services of the escapees, they would be put back to work immediately.[38] In late 1891, six labourers under advances ran away from the colliery and all were caught. Absconders usually escaped in small groups, indicating that such plans were premeditated. Most absconders were eventually caught.

In an Iban- and Malay-populated district, their physical appearance and distinctive dress made them stand out, and it was not easy for the Chinese to blend in with the local populace. Mine managers would secure the help of the local natives to trace the runaways. When five coolies fled from the mines in May 1917, 'word was at once sent to the *tua kampong* up and down river and Dayaks were told to be on the look-out. . . . On 20 May, these men were brought back by some Dayaks who had found them hiding in the jungle.'[39] Cash rewards of $5 were offered to those who turned in escapees.[40] If the Ibans were found to be sheltering runaways, they were fined.[41]

There were some instances of individual coal miners who took the risk of physically confronting the European mine managers. In 1908, the manager of the colliery, E. R. Naysmith, was killed by his cook:

> . . . Naysmith selected such labourers as he saw fit for domestic service in his house, where they came under the tyrannical supervision of his wife. The cook was himself picked in this way, and his repeated pleas to be allowed to return to his work in the mine fell on deaf ears. The murder occurred in the middle of a dinner party, and was motivated by the desire of the cook to forestall the execution on his own person of threats which had been uttered concerning the penalty which would be inflicted on him in retribution for the poor quality of the meal.[42]

This was of course a rare case of a miner killing a European. It was usually the workers who were on the receiving end of 'punishment' meted out. Europeans had to be treated with respect by the coolies.[43] The rattan whip was used by mine managers on labourers who had become 'disrespectful' or troublesome.[44] Charles Brooke, on at least two occasions, cautioned against the excessive use of the rattan. In 1895, in a letter to the mine manager, the Rajah advised:

I am in receipt of your letter dated 11th July 1895, making inquiry if you are to be allowed discretionary power in inflicting small punishments by *rattan* on your coolies. In the case you mention, where a coolie has proved himself to be a dangerous character and guilty of making frequent assaults with weapons, I think a few strokes inflicted at the time before the others may have a salutary effect—but endeavour to let this be the exception and not the rule—and I allow this authority to yourself only and not to your successors—as men's tempers under trying circumstances sometimes are apt to blind their judgement and no punishment should be inflicted in anger.[45]

One course of action undertaken by miners dissatisfied with their working and living conditions, and low wages, was to go on strike; this happened in June 1919. Telegraphic news was at once dispatched to Kuching appealing for help. Resident Beville Archer left immediately for the colliery, accompanied by forty-two armed constables. Fifty-three arrests were made. Seventeen men were sent to Kuching to stand trial and to be imprisoned, while the remaining thirty-six were punished with six strokes of the rattan each.[46] This was the only recorded instance of a mass strike. There could have been more strikes except for the presence of armed police and the possibility that the mine manager could always call upon the support of Malays and Ibans. The Brookes had learnt a useful lesson in the 1875 *kongsi*–Brooke war. Trouble with the Chinese could always be suppressed with the aid of Iban warriors.

In a belated token effort to improve working conditions and alleviate the problem of absconders, the colliery encouraged the use of a piece-rate system, putting on the contractors the onus of responsibility for making sure the labourers worked.[47]

Complaints about the malingering of mine coolies are continual, but it is not possible to find the habitual shirkers as owing to the contract system in vogue, the contractor is the only person who knows whether his men are working or not.[48]

However, the introduction of piece-rates apparently did not improve the situation:

In 1918, thirty coolies who had been recruited in and around Kuching, by *mandor* Ah Chong arrived here on the 17th August. They appeared to be

CHINESE LABOURERS AND THE STATE ECONOMY 189

quite hostile to their *mandor* for having promised them contract work on a new railway line, at excessively high rates, and other tempting promises. As soon as they arrived twelve of their number absolutely refused to do work of any kind offered to them.... [49]

There was little if any hope of coolies getting better working conditions. Most coolies who had arrived to work under the indentured system could not expect to achieve the dreams and hopes which had brought them to Sarawak. Many died before even being able to repay their passages and advance wages.[50] A brave plan to send a petition to the Chinese Protectorate in Singapore in 1905 floundered because of unsympathetic officials. The workers sent a representative to Singapore who returned disappointed; back in Kuching, he was arrested and imprisoned for his actions.[51] The failure of the coolies' dreams and aspirations can be gleaned from the preceding account of working and living conditions at the Sadong colliery.

Social Life

The majority of the labourers at the Sadong colliery were Cantonese and Hakkas, with smaller numbers of Teochius and Hainanese and, later, some Foochows. Like Chinese elsewhere in Sarawak and the *Nanyang*, a temple or 'joss-house' was central to the spiritual and emotional lives of these pioneers. A *Tua Pek Kong* shrine, deifying the pioneering spirit, was transferred to a new 'joss-house' in 1892.[52] *Wayang* shows were as popular here as in other places such as Simanggang or Marudi. The Sadong Chinese constructed a *wayang* house at the same time as the temple was rebuilt. Few of the men had Chinese or native wives with them.[53] Gambling[54] and opium-smoking[55] were virtually the only recreational pastimes available to the men, and while there may possibly have been prostitution as well, there is no evidence to show it existed. The gambling farm was open only on Saturdays, but, this did not deter the mine workers from gambling outside permitted hours.[56] Like the boat hawkers who hoped to get rich through gambling, the mine coolies harboured such dreams as well, and many gambled in the hope of escaping from the drudgery and wretchedness of their working and living conditions.

In the wider mining community, there was tension between the different Chinese dialect groups. Four hundred Hakka and Cantonese workers fought each other on 21 February 1901, and an extra police guard was called in from Kuching to maintain peace and order.[57] Two years earlier, thirty Hakka labourers had broken into the Simunjan shops, mostly Teochiu-owned, and beaten up the traders.[58] Dialect group rivalry among Chinese pioneering groups was not uncommon. In the coal mines, the miners spoke mutually unintelligible dialects, and this, plus the fact that Hakkas were considered to be inferior, accounted to some extent for the mutual resentment and hostility. Gambling losses and secret societies were also responsible for the rivalry between Hakkas and Cantonese. Secret societies controlled the gambling rackets. In February 1900, five *samseng* (gangsters), Soo Tee, Wong Wa, Fung Yow, Mok Seng, and Hee Hong, were charged with being members of an unlawful society: 'These highbinders were banded together for the purpose of demanding the winnings of Chinese gamblers with menaces and those not complying with their demands were threatened with assault or beaten, or otherwise molested.'[59]

Hakka and Cantonese ill-feeling towards the Teochius was due to indebtedness to bazaar shopkeepers who were mostly Teochius. In 1904, the Resident complained that the traders had been 'allowed to give a great deal too much credit to indentured coolies'.[60] As the wages of the labourers were insufficient, they were forced to buy food and other necessities on credit; they also borrowed money to pay their gambling debts.

Despite the tough working conditions and the sporadic violence on the mines, there was the very rare, happy occasion for the men to look forward to and look back on. Chinese New Year holidays provided welcome relief for the miners and, during this time, they would momentarily forget the agony of their work and celebrate the occasion in a carnival-like atmosphere of gaiety and merry-making. In 1898, before the Chinese lunar new year started, the coolies received their wages in advance and were given two days off:

During these holidays, several processions and bands paraded the village at the mines and then came down and spent the day at the bazaar acting, dancing and firing crackers. They ultimately came across to the government station and walked in procession with flags, all round the grounds with the *wayang* band in front. On arrival at the government bungalow, a number of crackers were discharged, they then proceeded to the hospital where the inmates were treated to a performance of the *wayang*; the crowd ultimately dispersed, returning to the bazaar across the river. This occurred twice, one day the *Kheh* [Hakka] procession and band performing, followed the next day by the *Macaos* [Cantonese]. For two days the place was quite *en fete* and very noisy.[61]

Wayang shows evoked cultural nostalgia and enthusiasm, and a sense of ethnic identification. That both Hakkas and Cantonese participated in the festivities on different days was indicative of their rivalry.

The Significance of the Sadong Colliery to the Brooke State

In terms of capital expended, coal mining was the most important economic activity other than planting which the Brookes invested in directly. The significance of coal mining to the Brookes challenges the commonly accepted notion that Brooke economic philosophy discouraged speculative capitalist enterprise. For the Brookes, coal became the single most important mineral export from 1889 to 1899. In the period from 1881 to 1899, the colliery consistently registered net profits, as can be seen from Table 9.1 below. Revenue figures did not take account of the coal used for internal consumption, particularly by local steamers, as this was delivered free of charge to the state. In other words, the state got the coal free of charge from the toil and sweat of Chinese labour.

In 1888, encouraged by the success of the Sadong colliery, the

TABLE 9.1
Sadong Coal Account, 1881–1899 ($)

Year	Excess of Expenditure over Revenue	Excess of Revenue over Expenditure
1881	4,030.57	–
1882	8,059.38	–
1883	24.85	–
1884	–	909.89
1885	–	4,377.93
1886	–	5,147.51
1887	–	6,273.55
1888	11,805.98	–
1889	5,505.08	–
1890	–	4,049.05
1891	4,418.68	–
1892	–	5,527.55
1893	510.17	–
1894	–	8,217.66
1895	5,082.40	–
1896	–	4,760.35
1897	–	41,789.80
1898	–	4,437.19
1899	–	919.59
Total	$ 39,437.11	$ 86,410.07

Source: SG, 2 July 1900, p. 132.

government purchased the rights to another coal mine, on Muara island at the mouth of the Brunei River, from an individual lessee, Cowie, who had originally obtained the lease from the Sultan of Brunei. An optimistic note was sounded for the Muara (later renamed Brooketon) mines, and the Brooke administration initially hoped to produce 3,000 tons of coal per month.[62]

But the mine turned out to be a financial disaster and was finally closed down in 1925. Although 650,000 tons were extracted, most of it was sold at a loss in Singapore. The distance from Singapore, the lack of storage facilities, expensive machinery, and export duties imposed by the Colonial Office combined to make the coal mine an unprofitable venture.[63] The estimated loss to the Brookes for the thirty-three years of the colliery's operations was $1½ million.[64] With the financial loss incurred in the Brooketon mines, the Sadong colliery was no longer seen as a viable investment. A substitute fuel source, oil, had become available at Miri, in the north of Sarawak. There were also recurring problems in recruiting Chinese labour for the hazardous underground work. Chinese *sinkheh* could find easier and more financially attractive work in rubber tapping. Labour shortages led to further decreases in output. On 16 May 1927, a *Sarawak Gazette* report stated:

> The [Sadong] colliery was unable to recruit any labourers, either at Singapore or Hong Kong, in place of those who had taken their discharge . . . the labour force had been depleted so seriously by the end of the year that output had fallen short of the local demand, and the Sarawak Steamship vessels had been obliged to take in bunker coal at Singapore as a result.[65]

When the Sadong mines were finally closed in January 1932, a labour force of 195 Chinese coolies remained. Less than half of the workers left for Kuching, and the rest stayed behind, taking up local employment and planting.[66] Simunjan, headquarters of the Sadong district, suffered a demise with the closure of the mines. The government headquarters was transferred to Serian, further up the Sadong River but nearer to Kuching by land.[67]

The Opening of the Oilfields in Miri

Oil had long been known to the natives of the Miri district, oozing from seepages in the ground. The oil, locally referred to as *minyak tanah* (earth oil), was used with resin to caulk boats and with open wicks for lighting fires. A Brooke official, Charles Hose,[68] who served in the Baram district after it was ceded to the Rajah in 1882, saw the

CHINESE LABOURERS AND THE STATE ECONOMY 193

potential of oil, and tried to persuade the ageing Rajah to drill the oil on a commercial basis. He was unsuccessful at first. It was only in 1907 that Charles Brooke finally relented, perhaps convinced that it was a commercially viable proposition. With the disappointing failure of the Brooketon coal mines and the decreasing output of the Sadong colliery, it had become expedient to search for an alternative source of fuel. Hose was authorized to open negotiations with the Anglo-Saxon Petroleum Company. The company was then granted a concession to drill for petroleum and other mineral products in Sarawak for a term of seventy-five years and was to pay a royalty of 1s for every ton extracted. An additional clause stated that the company should keep a minimum of 10,000 tons of liquid fuel in storage for the use of the British Navy.[69]

The oil company was thus confronted with the task of transforming the sleepy coastal village of Miri, comprising about twenty scattered Malay houses and two Chinese shops, into an oil-drilling industry. For this task, European capital and technology were to be utilized. After the exploratory and mapping work was completed, drilling began in 1910, in the vicinity of Miri village, 14 miles along the coast from the mouth of the Baram River. Being in a remote area, obtaining labour was a problem; like the Sadong colliery, the Anglo-Saxon Petroleum Company turned to Chinese and, to a lesser extent, native labour.

Labour Procurement and the Work-force Structure

The Anglo-Saxon Petroleum Company obtained its first batches of Chinese coolies from Singapore.[70] The men were probably recruited on similar terms of service as those hired for the Sadong colliery, with their passages and advance wages paid for by the company, and the men then being bonded to serve the company for an indefinite period.

A large pool of labour was needed to transport the equipment from ships anchored off-shore, hauling it along the ground, setting up oil derricks, digging wells, and drilling for the oil. Within a short space of three years, by 1913, 25,057 tons of oil had been produced from the wells.[71] By that time, an elaborate employment structure had evolved; this was pyramid-shaped, with 21 European managerial and technical staff at the top, followed by 6 'Asiatic' (origins unknown) clerks, 16 Chinese and 57 Malay artisans and, at the base, a pool of 411 labourers comprising 196 Chinese and 140 Iban coolies, 10 other Asiatics, and 65 labourers working for Chinese contractors. In other words, the company employed a total of 511 people.[72] In the beginning, there was an almost equal number of Chinese and indigenous labourers, but over the years, the number of Chinese workers increased at a faster rate than did the number of native employees. The economic and demographic pressures in south-eastern China that had forced Chinese men to go abroad to work were responsible for bringing many to the Miri oilfields. Figures for the labour force for the years 1914 to 1919 were not broken down by ethnicity, but the majority were Chinese. In 1914 there were 900 workers, increasing to 969 in 1915 and then decreasing slightly to 927 in 1916, and jumping to 1,062 in 1919.[73] By 1920, the labour force had almost doubled to 2,272; Table 9.2 shows the breakdown of employment by ethnic group in 1919 and 1920:

TABLE 9.2
The Miri Oilfields' Labour Force, 1919 and 1920

Labour Force by Ethnicity and Level of Skill	1919	1920
Chinese artisans	48	269
Chinese artisans on contract	–	3
Chinese coolies	740	983
Chinese coolies on contract	–	200
Malay artisans	135	298
Malay coolies	65	104
Javanese	–	178
Ibans	45	216
Indians	29	21
Total	1,062	2,272

Source: SGG, 16 March 1921, p. 68.

The phenomenal increase in employment in 1919 and 1920 was due primarily to increased production. In the war years from 1914 to 1919, the Anglo-Saxon Petroleum Company supplied 287,000 tons of oil to the British Navy.[74] In 1914, a submarine pipeline was laid from Tanjong Lutong, 7 miles south of Miri town, when increased production no longer warranted the hazardous and slow method of transporting oil by lighters to off-shore tankers. With a new submarine pipeline, it was considered necessary to construct a refinery in Lutong in 1917. By 1921, there were 101 wells, 10 new derricks and rigs, and annual production reached 141,492 tons.[75] More demands were placed for labour recruitment. In April 1919, the

Sarawak Gazette reported: 'Labour has been a difficult problem throughout the year, and at no time have we been able to import sufficient for our needs; repeated attempts to import sinkheh coolies from Singapore were unsuccessful and we have been compelled to try Hong Kong.'[76]

With Hong Kong as an additional labour source, there did not seem to be any further difficulties as a combination of chain migration, clan sponsorship, and local contractors directly going to Hong Kong to recruit coolies helped to increase the supply of labourers.[77] A census of workers taken in 1923 revealed a total of 3,860 coolies excluding Chinese contractors' employees, who numbered 754.[78] Unfortunately, there is no indication of the ethnicity of the coolies, but undoubtedly the majority were Chinese. The Chinese labour force was large enough to constitute a community by itself—one which was stratified according to the skills and jobs performed.

Working Life and Social Conditions

This account relies heavily upon oral interviews as there is little documented material available on work and life in the oilfields. Dangerous working conditions were omnipresent for the field labourers. A common danger was oil-wells flaring up suddenly when drilling for oil was being carried out. It was reported: 'Very often, the discovery of oil was heralded not by the gusher of movie fame but by a column of fire flaring out of the hole in the ground.'[79] In one incident in May 1927, a European supervisor and four Chinese coolies were killed when a fire broke out at a well.[80] Drilling was an occupational hazard as no protective aids were worn. A Cantonese labourer, Loh Kok Tong, a member of a well-pulling crew from 1917 to 1924, lost the skin from his hands and arms on seven occasions.[81]

Apart from the dangers faced at work, the coolies were plagued by diseases because of their poor living environment and inadequate nutrition. In 1923, it was reported that one half of labourers' deaths could be attributed to malaria.[82] Insanitary living conditions created an excellent breeding ground for the carrier of the disease, the Anopheles mosquito. The poor quality of food given to the coolies was responsible for outbreaks of beriberi. Poor working and living conditions were linked directly to social stratification among the employees.

The most degrading form of labouring work was performed by the tukang ayer ('t.a.'), which literally means 'water-drawer', and these men were considered to be little better than common servants. They did menial tasks like collecting firewood, cleaning living quarters, carrying water, and other miscellaneous, generally filthy jobs. They received only $0.45 a day on average, with a maximum of $0.55,[83] and had little hope of social and occupational advancement.[84] Other workers received daily wages ranging from $0.60 to $0.80.[85] These labourers were housed in 'coolie lines', of which there were eighteen in 1930, each with twelve dormitories.[86] Rice rations, salted fish, and cooking oil were supplied free once a week by the company.[87]

Gambling was a favourite pastime for an isolated community of men who, in their spare time, were often bored and who hoped, by gambling, to win their fortune with one stroke of luck. There were no clubs or associations and hence no social outlets for the men, and this was one reason why most of the men turned to gambling:

> Every pay day, the whole town would suddenly take on the appearance of a fair. From the hidden depths of the little houses along the coolie lines emerged tables of all shapes and sizes for all kinds of gambling games from *fan tan* to *chap ji ki*. Eating stalls offering culinary delights to titillate the most jaded palate would materialise as if by magic to cater to the needs of obsessive gamblers.[88]

According to informants, gambling was common even on days other than pay day, and on Sundays.[89] Incidences of suicide by hanging outside the 'coolie lines' were reported to be a common sight, averaging one to two a month, particularly of those who had run into debt.[90]

Immediately above the daily-wage labourers in the worker hierarchy were the artisans. They normally had some kind of skill. One such artisan was Chan Wey Yuen,[91] a Shanghainese, who arrived in Miri in 1930 at the age of thirty. He had been in Singapore for some eight years prior to his travelling to Miri, where kinsmen and friends had asked him to join them. Upon arrival, he entered the services of the company, and was put in charge of a group of men doing woodwork, making furniture, and constructing and repairing houses. Chan received wages higher than the labourers, getting $1.80 a day.[92] Artisans received better accommodation than the labourers, and this informant (Chan) shared artisans' quarters with three others. They had the additional privilege of catering services in the mess, for which they paid $12 per month. The men had their own association, the Chinese Engineering Association, a reflection of their technical skills and qualifications, as well as the better working conditions and higher social status they enjoyed as compared with the labourers.

At a higher level of occupational status were the 'senior Asiatic staff', the clerks who performed the administrative and paper work for the company. Typical of this category of the staff was Lee Yok Seng, a Kuala Lumpur-born Cantonese, who arrived in Miri in 1933 at the age of nineteen.[93] Lee started off as a clerk in the machine shop, receiving a monthly salary of $40, with free accommodation provided. 'Senior Asiatic staff' members, like the tradesmen, had their own rooms. They received additional privileges such as free crockery, bedsheets, pillows and mattresses, and a *tukang ayer* was assigned to each staff block. Clerks like Lee, most of them local-born, had received some form of English education, usually at missionary schools.

Having been exposed to Western education and values,[94] the company clerks aspired to social activities enjoyed by members of the élite European community.[95] The clerks had their own Miri Recreation Club, and played traditionally Western games like football, tennis, and basketball. After the Europeans had formed the Miri Rowing Club, the clerks acquired a couple of *sampan* and formed the Senior Asiatic Boat Club.[96] Social segregation was strong in Miri: if English-educated Asian employees harboured any hopes of being accepted socially by the Europeans, they were disappointed. The Gymkhana Club, for example, was an exclusive European social and sporting body that opened its doors to Asian members only after the Second World War. The Chinese employees at the Miri oilfield were very conscious of their inferior social position. According to Chan Wey Yuen, 'The Europeans generally looked down on the Chinese; this did not seem to bother the workers, who, being poor, would rather keep their tongues tied.'[97]

Labourers and Strikes

Small groups of men would occasionally try to escape from the harsh conditions of their workplace. Like the Sadong colliery, the Anglo-Saxon Petroleum Company relied upon sympathetic Malays and Ibans to round up the escapees.[98] When caught, the escapees received the usual punishment of flogging and imprisonment. Individual acts of defiance by workers were rare, but group action was more common.

Employee solidarity was strong, and the coolies were prepared to act together, support each other, help those in trouble, and fight against management injustice. In June 1911, sixty Chinese coolies demonstrated outside the Miri Court-house and attempted to rescue three fellow prisoners. Failing to do so, they proceeded to the *mandor*'s quarters and threatened him, but were stopped before they could carry out their threats. Acting Resident H. S. B. Johnson, on hearing news of the 1911 strike, proceeded to Miri from Marudi (which was then the divisional capital), arrested five leaders amongst the protestors, and sentenced each of them to seven years' imprisonment with hard labour.[99]

In March 1914, a secret society was formed 'for the purpose of helping and standing by one another when in trouble ... raising money, and meetings were being held somewhere in the jungle'.[100] The secret organization held its meetings with such efficiency that attempts by the company to break up these meetings never succeeded. Eventually, six of its leaders were caught and sent to Kuching to be imprisoned. However, this did not mean the end of labour unrest. Relations between the company management and the coolies were at breaking point by October 1914. Three labourers were arrested for using threatening language in front of a company official, and they were duly sentenced to six months' imprisonment. The whole workforce went on strike for four days; in view of the volatile situation, persuasion rather than force had to be used to persuade the coolies to return to work. The ringleaders of the strike were banished from the country.[101] Whenever there was a strike, the leaders would be identified and imprisoned, or banished from the country. The company refused to consider the basic causes of industrial strife. It consistently underestimated the strength of worker discontent, failing to see that the workers were earnest in their demands for better pay and safer working conditions.

Again in March 1922, the coolies went on strike, demanding justice for the death of one of their fellow workers and improved working conditions. Efforts by leading Miri Chinese merchants to mediate, together with an agreement on the part of the management to meet the workers to discuss their demands, brought the situation back to normal.[102] In July 1923, the Chinese labourers asserted their demands even more forcefully, and held a demonstration. The action precipitating the demonstration was a case of assault by a Javanese on a Chinese, a non-company employee. The occasion was used by the coolies to express their grievances against the company. Retribution was demanded by the company workers who gathered in front of the police station, and the situation got out of hand when groups of Chinese began to throw stones at the station. Owen, the Resident, ordered the police to open fire: the result was 13 persons dead and 24 wounded. An August 1923 report in the *Sarawak Gazette* tacitly recognized that the workers were within their rights when they demanded legal action against the assailant, but chose to interpret the situation as an occasion 'for the rough Chinese element to assert its disregard for law and order'.[103] A Commission of Inquiry investigating the incident exonerated Owen of any blame and defended his actions instead. Nothing further was heard about industrial action and the demands of the coolies; this was due to the harsh coercive measures taken by the Brooke administration to curb industrial unrest, to the falling quality of reporting in the *Sarawak Gazette*, and to stricter censorship being practised, rather than to acquiescence and a 'towing of the line' on the part of the labourers.

Chinese coolies on the coal mines and oilfields contributed significantly to the economic development of Sarawak by providing cheap labour. The work was tough, dirty, and dangerous, and the wages were low. The indentured system of recruiting labour was almost akin to slavery, and it offered the labourers little hope of improving their material standard of living.

1. An example of this view can be found in S. Baring-Gould and C. A. Bampfylde, *A History of Sarawak under Its Two White Rajahs 1839–1908* (London: Sotheran, 1909; reprinted Singapore: Oxford University Press, 1989), p. 417.

2. Charles Brooke put forward this poser in 1907: 'Did it ever strike those in power that the object of a government should be peace, happiness and contentment to all, and especially to the natives of the soil?' (Charles Brooke, *Queries: Past, Present and Future* (London: The Planet, 1907), p. 14.)

The notion of the Brookes justifying their rule because of their belief in the idea of trusteeship of native interests still persists among present-day writers. See C. M. Turnbull, *A Short History of Malaysia, Singapore and Brunei* (Singapore: Graham Brash, 1980), p. 160.

3. In the late 1880s, Charles Brooke lost $60,000 after a tobacco-planting venture in Lundu failed. See Letters of Rajah Charles Brooke, Vol. 2, 20 November 1890, p. 120, SA. The second Rajah also invested $176,000 in coffee and tea estates on Matang Mountain and Setap near Kuching at the turn of the century. See *SG*, 1 April 1902, p. 14.

4. Spenser St. John, *The Life of Sir James Brooke, Rajah of Sarawak* (London: William Blackwood & Sons, 1879), p. 139.

5. Colonel Orfeur Cavenagh, *Report upon the Settlement of Sarawak* (Calcutta: J. Kingham, 1863), p. 13.

6. *SG*, 31 January 1872, p. 10.

7. *SG*, 16 September 1872, pp. 70–2.

8. *SG*, 31 March 1873, p. 28.

9. *SG*, 16 May 1874, p. 3.

10. *SG*, 2 November 1886, pp. 104–5.

11. *SG*, 1 July 1886, p. 107.

12. *SG*, 1 August 1882, p. 56. See also entry dated 19 July 1883 in The Rajah's Order Book, 1863–1890, SA.

13. *SG*, 1 November 1882, pp. 133–4.
14. Cited by Craig Lockard, 'Chinese Immigration and Society in Sarawak, 1868–1917' (MA thesis, University of Hawaii, 1967), p. 29.
15. *SG*, 1 August 1888, p. 82.
16. *SG*, 1 June 1888, p. 72.
17. *SG*, 1 June 1891, p. 90.
18. Letters of Rajah Charles Brooke, 14 February 1895, p. 231, SA.
19. *SG*, 2 March 1896, p. 50.
20. *SG*, 4 August 1903, p. 158.
21. 'Recollections of an Ex-Government Officer, Haji Ahmad Zaidell bin Haji Tahir', *SG*, 31 January 1972, pp. 6–8; Chong Ah Onn, 'Life among the Primitive People of Borneo', personal papers, 1960, p. 59, SA.
22. *SG*, 1 May 1894, p. 70.
23. *SG*, 3 October 1906, p. 237; and Letters of Rajah Charles Brooke, Vol. 10, 1 September 1906, p. 187, SA.
24. *SG*, 1 June 1889, p. 86.
25. *SG*, 1 August 1891, pp. 121–2.
26. *SG*, 17 March 1924, p. 150.
27. *SG*, 2 November 1891, p. 163.
28. *SG*, 1 February 1892, p. 185.
29. *SG*, 1 March 1894, p. 42.
30. *SG*, 1 April 1897, p. 69.
31. *SG*, 1 April 1899, p. 71.
32. *SG*, 1 October 1898, p. 195.
33. *SG*, 1 June 1889, p. 119.
34. *SG*, 1 May 1922, p. 128.
35. Letters of Rajah Charles Brooke, Vol. 10, 23 November 1904, pp. 17–18.
36. K. H. Digby, *Lawyer in the Wilderness*, Cornell University, Southeast Asia Program Data Paper No. 114 (Ithaca, New York: Cornell University Press, 1980), pp. 60–1.
37. Ibid.
38. *SG*, 1 June 1891, p. 90.
39. *SG*, 2 July 1917, p. 163.
40. *SG*, 1 June 1898, p. 124.
41. *SG*, 1 June 1922, p. 153.
42. Digby, op. cit., p. 61; and *SG*, 1 August 1908, p. 196.
43. *SG*, 1 May 1891, p. 70.
44. Letters of Rajah Charles Brooke, Vol. 4, 10 March 1898, p. 242, SA.
45. Letters of Rajah Charles Brooke, Vol. 3, 18 July 1895, p. 298, SA.
46. *SG*, 16 July 1919, p. 193.
47. *SG*, 5 August 1907, p. 195.
48. Ibid.
49. *SG*, 16 October 1918, p. 270.
50. *SG*, 1 May 1889, p. 68.
51. Letters of Rajah Charles Brooke, Vol. 14, 12 December 1905, p. 58, SA.
52. *SG*, 2 January 1892, pp. 11–12.
53. *SG*, 1 April 1932, pp. 71–2.
54. *SG*, 2 November 1907, p. 251.
55. *SG*, 1 April 1927, p. 92.

56. *SG*, 2 November 1907, p. 252.
57. *SG*, 1 April 1901, p. 83.
58. *SG*, 1 February 1899, p. 21.
59. *SG*, 1 May 1900, p. 98.
60. Letters of Rajah Charles Brooke, Vol. 10, 29 October 1905, p. 117, SA.
61. *SG*, 1 April 1899, p. 171.
62. *SG*, 1 October 1888, p. 122.
63. *SG*, 12 January 1905, p. 122; and Baring-Gould and Bampfylde, op. cit., p. 357.
64. *SG*, 1 May 1925, p. 95.
65. *SGG*, 16 May 1927, pp. 246-7.
66. *SG*, 1 April 1932, pp. 71-2.
67. *SG*, 1 February 1936, pp. 33-4.
68. Charles Hose, 'The Metamorphosis of Miri', *British Malaya*, No. 11 (June 1927), p. 41.
69. Agreement Book, Vol. 3, January 1902-July 1909, pp. 210-11, SA.
70. Entry dated 20 April 1910 in The Rajah's Order Book, Vol. 3, p. 371, SA.
71. G. C. Harper, 'The Miri Oil Field, 1910-1972', *SMJ*, Vol. 20, No. 40-1 (January-December 1972), p. 26.
72. *SGG*, 1 March 1913, p. 21.
73. *SGG*, 16 February 1915, p. 43; 3 April 1916, p. 74; 16 March 1917, pp. 59-60; 17 May 1920, p. 173.
74. Sarawak Shell Berhad, *The Miri Story, the Founding Years of the Malaysian Oil Industry in Sarawak* (Miri: Sarawak Shell Berhad, 1979), p. 16.
75. *SG*, 1 April 1921, p. 47.
76. *SG*, 1 April 1919, p. 97.
77. This has been confirmed by interviews conducted in Miri. Interviews with Chew Peng Man, 8 August 1981, and Chan Wey Yuen, 10 August 1981, both in Miri.
78. *SGG*, 16 May 1923, p. 174.
79. Sarawak Shell Berhad, op. cit., p. 21.
80. *SG*, 1 August 1927, p. 123.
81. 'The Saga of Slim', *Salam*, 19 February 1955, p. 3.
82. *SG*, 1 April 1924, p. 173.
83. Information from Chai Shun Fatt, 15 August 1981, Miri. Chai joined the oil company as a mail clerk in 1936.
84. There was the rare case of a *tukang ayer*, Yong Ah Fook, who joined the company in 1932 and continued in this position until 1956, when he was promoted to be a painter, a position he held until his retirement in 1966. (*Salam*, 30 June 1966, p. 4.)
85. *Salam*, 15 January 1955, p. 1; and interview with Leong Kam Fook, 7 August 1981, Lutong.
86. Information from Chan Wey Yuen, 10 August 1981, Miri.
87. Information from Kong Pit Leong, 12 August 1981, and John Leong, 7 August 1981, both of Miri.
88. Sarawak Shell Berhad, op. cit., p. 27.
89. Information from Chai Shun Fatt and Chu Nyin Hee, both interviewed on 13 August 1981, Miri.
90. According to information given by Chai Shun Fatt, 13 August 1981, Miri.
91. See note 77.

92. As a comparison, masons working for the Public Works Department in Kuching received between $1.20 and $2.00 per day. Entry dated 30 January 1912 in The Rajah's Order Book, Vol. 3, p. 292, SA.
93. Interview with Lee Yok Seng, 26 July 1981, Miri.
94. Examples would be interviewees like Chai Shun Fatt (note 83), born in Sabah, but educated in St. Thomas' School in Kuching, and Ho Hong Kwong who attended a mission school in Marudi.
95. Miri had a large European community, even surpassing that of Kuching: in 1921, there were 131 European men, women, and children in the town. The Europeans lived in bungalows in Tanjong Lobang, Lutong, and Miri. They had their own clubs and social activities and had little social contact with the local populace.
96. Sarawak Shell Berhad, op. cit., p. 28.
97. Interview with Chan Wey Yuen, 10 August 1981, Miri.
98. *SG*, 16 January 1915, p. 23; and 2 March 1914, p. 58.
99. *SG*, 1 August 1911, p. 152.
100. *SG*, 2 March 1914, p. 58.
101. *SGG*, 16 February 1915, pp. 39–40.
102. *SG*, 1 May 1922, p. 130.
103. *SG*, 1 August 1923, p. 226; and 3 September 1923, p. 277.

1 *Kongsi*-house of the 'Fifteen Company' in Marup, near Engkilili.

2 Interior of the *kongsi*-house of the 'Fifteen Company'.

3 Flag-poles symbolizing the autonomy of the 'Twelve Company' in Bau. According to oral tradition, the *kongsi* flag was raised here every morning. *Kongsi* members sentenced to death were executed here in front of the flag.

4 A late nineteenth-century photograph of Chinese miners in a gold mine. (Courtesy Sarawak Museum.)

5 A temple built in honour of Liew Shan Pang, leader of the 'Twelve Company' in Bau. This temple was rebuilt in 1976. It marks the spot where Liew was supposed to have been buried.

6 Chinese traders buying camphor from Kayans. (From W. H. Furness, *The Home Life of Borneo Headhunters, Its Festivals and Folklore*, Philadelphia: Lippincott, 1902.)

7 Ibans bargaining over the sale of valuable Chinese jars. (From W. H. Furness, *The Home Life of Borneo Headhunters, Its Festivals and Folklore*, Philadelphia: Lippincott, 1902.)

8 Ibans on a trip to a bazaar. Contact with Chinese traders has brought about changes in material culture for the Ibans. The man on the right is wearing a fez and a Chinese shirt. The woman who is standing is wearing a skirt made of black *blachu*. (From R. H. W. Reece, *The Name of Brooke: The End of White Rajah Rule in Sarawak*, Kuala Lumpur: Oxford University Press, 1981.)

9 Iban clients together with Chinese traders in a longhouse. (Courtesy Sarawak Museum).

10 A Chinese boat hawker in front of his boat, 1981.

11 A late nineteenth-century view of the Kuching waterfront. (Courtesy Sarawak Museum.)

12 A longhouse. (Courtesy Sarawak Museum.)

13 A Chinese trader in Belaga, 1981.

14 A rattan raft being floated down to a bazaar. (From W. H. Furness, *The Home Life of Borneo Headhunters, Its Festivals and Folklore*, Philadelphia: Lippincott, 1902.)

15 Marudi bazaar, 1902. (From W. H. Furness, *The Home Life of Borneo Headhunters, Its Festivals and Folklore*, Philadelphia: Lippincott, 1902.)

16 Chinese ceramics. (Courtesy Sarawak Museum.)

17 Sing Chio Ang (Sungei Merah), Sibu, 1903. Sungei Merah was one of the two sites for the first batch of Foochow settlers. (Courtesy Ling Kai Cheng.)

18 The second home of Reverend and Mrs Hoover in Island Road, Sibu, 1905–8. (Courtesy Ling Kai Cheng.)

9 The first Anglo-Chinese School in Sibu, 1903. (Courtesy Ling Kai Cheng.)

10 The first quarterly Methodist Conference, 1903. (Courtesy Ling Kai Cheng.)

21 The first Methodist chapel in Sibu at Sing Chio Ang (Sungei Merah), 1903. (Courtesy Ling Kai Cheng.)

22 A rubber smallholding. (Courtesy Sarawak Museum.)

10
The Chinese and the Brookes

WHEN James Brooke acquired Sarawak, an uninviting piece of territory in Borneo, he had ambitious plans of making that state a part of an extended island empire, stretching from Borneo to Java and linked with the British colonies of Australia and New Zealand. He had clear objectives in mind for Sarawak. The Rajah hoped to encourage trade and economic development; he hoped to persuade Britain to take over Sarawak as a Crown Colony, thereby securing his own political future; he also saw himself as having a 'civilizing' mission in reforming native character. He said: 'I wish to correct the native character, to gain and hold an influence in Borneo proper, to introduce a better system of government, to open the interior, to encourage the poor natives, to remove the clogs on trade, and to develop new sources of commerce.'[1] For the daunting task of opening up Sarawak, James Brooke looked to the Chinese, whom he hoped would be the pioneers to dig the mines, farm the land, and trade with the natives.

The presence of the Chinese during the early years of the Brooke state was considered indispensable because of their contributions to the Brooke treasury. A dependent economic relationship between the Chinese and the Brookes thus evolved. But this economic role for the Chinese also presented the Brookes with a dilemma: the Rajahs wished to encourage the useful presence of the Chinese but, at the same time, they wanted to assert their sovereignty over them.

Indirect Brooke Rule over the Chinese

The most visible impact of Brooke rule over the Chinese was in the capital, Kuching; however, a detailed discussion on this subject lies outside the scope of this book. Nevertheless, Brooke rule in Kuching will be discussed here for it reveals much about Brooke attitudes to the Chinese in general, and it had direct implications for the way the Chinese outside Kuching were being governed.

It is not known how James Brooke dealt with the small community of Hokkien and Teochiu traders in Kuching, but it can be imagined that they were on amicable terms with the Rajah, for they took no part in the *kongsi*–Brooke war of 1857. The second Rajah developed close ties with the paramount Chinese leaders of the 1870s and 1880s—with the Teochiu leader, Law Kian Huat,[2] the Chao-ann leader, Chan Ah Koh, and the Hokkien leader, Ong Ewe Hai.[3] The Hokkien, Chao-ann, and Teochiu trading communities in Kuching had risen to economic prominence after the decline of the Hakka *kongsi* in Bau, and their commercial networks extended all over Sarawak. The community leaders of these groups were wealthy *towkay*, the ramifications of their economic power being felt far beyond the capital. It was possibly before the end of the nineteenth century that the institution of the Kapitan and Kapitan China General was set up. *Kapitan* were appointed for each of the communities, the positions going to wealthy *towkay*. *Kapitan* acted as intermediaries between government and people, and as unpaid advisers to the Rajah on economic, cultural, and social matters, and in various capacities as coroners, 'magistrates', and representatives in the Bankruptcy and Debtors' Courts.

The closest that the Chinese community in Kuching came to being self-governing in non-political matters was from 1911 to 1920, when a Chinese Court was instituted.[4] It was Charles Brooke's intention that the growing Chinese community in Kuching, and in Sarawak as a whole, be given direct autonomy in non-governmental affairs. Ong Tiang Swee, who was Kapitan China General, headed the court together with six other deputy magistrates, three of them representing the Hokkien community (which also served the Chao-ann, Henghua, and Foochow groups), and one each for the Teochius, Cantonese, and Hakkas. All the magistrates were well-known traders. The Chinese Court had jurisdiction throughout Sarawak. It involved itself in a wide range of matters, from Chinese customs, marital differences, division of property, partnership disputes, and investigation of bankruptcies, to appeals from the Debtors' Court, and the registration of marriages and betrothals in Kuching as well as outside the capital. The court was dissolved in 1920 on the grounds that Ong Tiang Swee, the Kapitan China General, was too preoccupied with his business interests.[5] It was not inconceivable that the court had become more powerful than was originally intended, having usurped the powers formerly exercised by the European Residents and District Officers.[6] In any case, the institution was a novel and short-lived attempt at delegating some measurement of self-government to the Chinese.

With the demise of the Chinese Court, the *kapitan* gradually lost whatever direct influence they might have had with the government. The third Rajah, in 1929, established a Secretariat for Chinese Affairs, manned by Chinese-speaking English officers, to take over some of the legal functions of the court, such as presiding over business and matrimonial disputes, and matters involving Chinese customs. The growing size and complexity of the Chinese community throughout Sarawak forced the Secretariat to take on other functions, such as the licensing of Chinese schools, monitoring Chinese political activities, and acting as the Protector of Labour, Women and Children. The Secretariat informally co-opted the services of the various Kuching *kapitan* and other influential persons, relying upon their advice.

Having looked briefly at *kapitan* administration and the nature of indirect rule in Kuching through the Chinese Court, and seeing how, as institutions, both were losing their effectiveness due to government fear of their potential power and to the encroachment of government bureaucracy, indirect rule over the rural Chinese will now be examined. While the *kapitan* in Kuching found their self-governing powers gradually being eroded away, *kapitan* in the outstations did not have any real power to exercise in the first place.

Beyond Kuching, there were two forms of indirect rule, through the *kanchu* and the *kapitan*, depending on the locality. Though different in name, both performed similar functions. In rural areas around Kuching, where large numbers of Chinese worked as cultivators and labourers, the *kanchu* was the headman.[7] The *kanchu* was the intermediary between the Brookes and the Chinese. He monopolized the sale of pork, arrack, and opium, and controlled pawnbroking and gambling. In addition, he was responsible for maintaining public order and had limited judicial power to handle minor civil cases. In 1886, the power of the *kanchu* was effectively circumscribed. He performed a perfunctory role in inspecting the pepper and gambier gardens in his district and weighing the produce sent out.[8] The bazaars of the Lupar, Rejang, and Baram Rivers, as well as the Foochow-populated districts of the Rejang, had prominent *towkay* and planters holding the office of *kapitan* and *kanchu* respectively. Bazaar *kapitan* only first appeared towards the end of the nineteenth century, and Foochow *kanchu* a few years after the Foochow colony was founded. *Kapitan* in the bazaars performed symbolic functions, acting as registrars of Chinese marriages and betrothals, and they were conveyors of government news to their respective communities.

The very personal and somewhat arbitrary style of rule of the European Residents and District Officers, aided by their subordinates, the Native Officers, and by the Chinese and indigenous court writers, did not give the bazaar-based *kapitan* much leverage to exert in their own communities, except in the area of Chinese customs, marriages and betrothals, and domestic disputes. If the bazaar communities were small enough and near enough to a Brooke station, symbolized by the fort, the Brooke officers and their staff could intervene directly in community affairs. Chinese court writer Cheyne Ah Fook, who was in sole charge of the Krian post at Kabong, personally mediated in a group fight between Chinese and Malays in December 1906.[9]

The *kapitan* were generally lacking in formal power as leaders, and they themselves were at times reprimanded for offences such as using false weights and contravening the rules for the running of the opium, arrack, and gambling farms. In 1896, a prominent shop in Marudi, Kim Wat, owned by a future *kapitan*, Lim Cheng Soon, was fined $6 for being in possession of a fake measure.[10] In Simanggang, Seng Joo, a shop belonging to the *kapitan*, was fined $25 in 1913 for 'recooking and selling opium dross'.[11]

But it was rare for a prominent *towkay* in Kuching to be slighted. The Police Superintendent, C. W. Daubeny, was cautioned in 1891 against allowing his men to be too rough in dealing with the Chinese in Kuching.[12] The Rajah was dependent on the wealthy Kuching *towkay* who monopolized the lucrative farm tenders to provide a substantial revenue to the Brooke treasury, and a show of goodwill towards them was considered vital and necessary.[13]

The Foochows were indirectly ruled through district headmen, whose powers and functions were similar to those of the *kanchu* outside Kuching in the nineteenth century, with the exception that the usual gambling, drinking, and opium farm monopolies did not exist in these Christian settlements due to a prior arrangement made between the first Foochow 'proprietor', Wong Nai Siong, and Rajah Charles. At the start of the Foochow colony, as was commonplace with other centres and areas in Sarawak, only prominent planters and entrepreneurs like Lau Ka Too and Wong Gin Huo[14] were recognized as *kanchu*, together with Reverend Hoover, the government-appointed 'Protector of the Foochow'. The Foochow *kanchu* performed similar functions as the bazaar *kapitan*, settling personal disputes, registering births, deaths, and marriages, and acting as intermediaries between their own people and the Brookes. For example, in 1927, when migration cuts were proposed, the *kanchu* of each area were consulted,[15] and in 1930, the Secretary for Chinese Affairs met with them to discuss issues such as unemployment and financial assistance to plant padi.[16] The Foochow *kanchu* enjoyed more autonomy than the bazaar *kapitan*. This was primarily due to the fact that infrequent visits were made by the Rajah's officers to the settlements which were dispersed along the banks of the Rejang River and were somewhat inaccessible. The *kanchu* were far less effective in the area of ethnic relations, playing no part in resolving land disputes in the Rejang, and relying instead on the government for help.

The only real powers which the *kapitan* and *kanchu* could boast of were advising the government on Chinese customary law and acting as registrars. The offices of *kapitan* and *kanchu* were retained for purely symbolic reasons. In the 1930s, when the haphazard legal structure in Sarawak was being rationalized in stages, Chinese headmen both in and beyond Kuching lost even more of their ascribed powers in dealing with their respective communities. Cases of customary law which previously were heard by the *kapitan* and *kanchu* now came before the formal legal courts.

The Chinese and the Legal Structure

James Brooke enacted the first set of laws for Sarawak in 1842, and it consisted of eight simple clauses on the maintenance of law and order, but did not deal directly with the Chinese. The Rajah was careful to enshrine his legal principles under the guise of the *undang-undang* (Malay law code) but vested the ultimate sovereignty and source of law in himself: 'The Governor issues these commands, and will enforce obedience to them; and whilst he gives all protection and assistance to the persons who act rightly, he will not fail to punish those who seek to disturb the public peace, or commit crimes....'[17]

The law, as it was interpreted in Sarawak, was summary and arbitrary. The source and interpretation of the law stemmed from the Brookes themselves who applied their own particular English principles of fair play and morality on their subjects.

There was no clear indication of how James Brooke would apply his own sense of justice to the Chinese, but of utmost concern to him was the question of his sovereignty, acknowledgement of which he was at pains to exact from the Ibans and the Chinese, and from enemies whom he was to variously label as 'pirates' and 'rebels'. Right from the start of the transfer of sovereignty from the Sultan of Brunei to the Rajah, James Brooke and the Bau Hakkas clashed over legal differences of opinion concerning the *kongsi*'s mining rights and rights of criminal jurisdiction.[18] The legal differences were, of course, part of a wider *kongsi*–Brooke political and economic rivalry, but such differences were important when considering the question of whether the rule of *kongsi* law or the rule of the English Rajah should reign supreme.

A good indication of James Brooke's views on justice is contained in a letter which he wrote in 1847, when he was back in England, to three magistrates whom he had appointed to form a board in his absence:

> The magistrates are not bound to administer justice according to the letter of the English law, but are to judge in equity between man and man, upon the broad and immutable principles of justice. ... To such a court the multiplied forms, rules, motions, notices and technicalities of British law would be a serious detriment or hindrance; in rejecting all such, as inapplicable and useless clogs, they will retain the grand foundation of British law, which is the eternal principle of justice.[19]

James Brooke, as an Englishman, thus selectively imposed English legal standards and values upon Sarawak, a transformed legal system devoid of checks or redress. The rules of justice were left to the discretion of the European officers, guided only by the broad outlines laid down by the Rajah. This tradition of discretion and informality continued during the long reign of Charles Brooke who, at the same time, issued a number of orders, rules, and regulations, some of which directly concerned the Chinese, particularly with respect to interethnic relations. The laws (called 'Orders') are contained in the Rajah's Order Books, begun in 1863 and the source of the present laws of Sarawak. In spite of the apparently arbitrary way in which court cases were handled, there was an elaborate legal structure established during the reign of Charles Brooke.

At the top of the judicial system was the Supreme Court, based in Kuching. It was presided over by the Rajah, the Resident in Kuching (the most senior Brooke officer), the Treasurer, and three Malay *Datu*, who represented the Malay élite. The Supreme Court had jurisdiction over all other courts, hearing appeals from the Resident's courts and confirming sentences of death and penalties which carried over 10 years' imprisonment. It was in the Supreme Court that the killers of murdered traders were tried, and where the leaders of Chinese secret societies were sentenced to death.

Below the Supreme Court stood the Police Court and the General Court. The Police and General Courts in Kuching, were headed by the Resident and one *Datu*. In the outstations, the Resident of that Division and at least one member of the largely symbolic Council Negri, usually a Native Officer, presided. The Police and General Courts were responsible for hearing cases of larceny, misdemeanour, and more minor offences such as fake weights, 'absconding under advances', and 'living in Dayak houses'. Fines and terms of imprisonment were limited to $30 and six months' jail.

Then there were the Debtors' Court and the Court of Requests. In Kuching, these were presided over by a magistrate. Outside the Brooke capital, these were headed by an Assistant Resident with one Chinese magistrate, usually a headman or a court writer (normally a Chinese or native employee). The jurisdiction of this court covered the settling of debts or claims arising out of the dissolution of partnerships, disputes over debtors, and debtors' accounts.

Of direct importance to the Chinese traders was the Bankruptcy Court, established in 1881, after Rajah Charles received a petition from Chinese and Indian merchants in Kuching complaining that debtors were dealt with too leniently in the Debtors' Courts and that bankruptcy was too easily granted. There was a Bankruptcy Court in each of the following towns—Kuching, Mukah, Sibu, and Simanggang, centres with clusters of Chinese traders. Each court was presided over by the Resident of the division, the Chief Native Officer, and three magistrates nominated by the Rajah for each of the districts.[20] The Bankruptcy Court served to safeguard the interests of Chinese and Indian creditors who gave out advances to small traders and native clients.

The different courts were in fact all part of a single institution headed by the Resident, performing different functions on different days. The Police and General Courts were the busiest, with the most number of cases heard by a Resident, from land disputes to thefts and fighting. Much power was vested in the Resident. Legal niceties, even when defined within the framework of the Rajah's Orders, were dispensed with, and Charles Brooke allowed his officers to make their own judgments on the spot according to their sense of justice, equity, and the prevailing circumstances. This system of meting out justice did give rise to complaints over the conduct of Residents. In 1910, Charles Brooke wrote a letter of reprimand to one of his officers in the coastal station of Oya:

I have received a very strongly worded petition signed by almost all Malays and Chinese in Oya against your administration of justice in that Residency. I shall have to make inquiry into the matter and it will be necessary to remove you as soon as possible as you have forfeited the goodwill and confidence of the inhabitants.[21]

Informality and force of personality were the touchstones around which Residents and their subordinate officers operated the legal system. The officers spent a lot of time travelling to remote bazaars and longhouses. One of their most time-consuming duties was arbitrating in court cases.[22] The hearing of court cases served as one of the principal means of contact between government and people. In the 1920s, however, this informal way of hearing court cases came under threat. It has been noted elsewhere that the establishment of the Chinese Court for a brief period between 1911 and 1920 was a novel experiment, for it gave due recognition to Chinese customary law. The abolition of the Chinese Court in 1920 was part and parcel of an attempt to rationalize the legal structure, measures for which were introduced in 1922.[23] The new court system consisted of two tiers comprising the Supreme Court and the Residents' Courts, which formed the Supreme Court and the District, Magistrate's and Native Courts, forming the Lower Courts. In 1927, an edition of Chinese family law was published by the government, after consultation with the various Chinese *kapitan*, but it does not seem to have ever been considered as an integral part of the body of the official customary law of Sarawak. The years between 1927 and 1941 were active ones for the rationalization of the legal system in Sarawak. In 1928, for the first time in Sarawak's history, a professionally trained legal officer, T. Stirling Boyd, was brought to Sarawak, and a new office of Judicial Commissioner was set up.

A significant enactment, which now emphasized in law what in the past had only been carried out in a discretionary fashion by the Rajah and the Residents, was the Law of Sarawak Order of 1928, with the following clause: 'The Law of England in so far as it is not modified by Orders and other Enactments issued by H.H. the Rajah of Sarawak or with his authority, and in so far as it is applicable to Sarawak having regard to native customs and local conditions, shall be the law of Sarawak.'[24]

Chinese customary law was now interpreted within the framework and principles of English law. In an example of a domestic dispute in 1936, *Kho Leng Guan* v. *Kho Eng Guan*, the brief of the case was as follows:

> The appellant, Kho Leng Guan, lived for some years with his mother and brother, and maintained a joint family establishment. The mother died in 1930 and the two brothers continued to live together. In October 1936, they had a quarrel and separated, and the respondent brought an action against his brother on the ground that by Chinese custom, he was entitled to a share in the family property.[25]

The respondent failed to win his case. The court's verdict was:

The relation between the parties is primarily a relation of contract, and as such, foreign law [Chinese customary law] does not in [the court's] view apply to a case in which the contract was made and performed in Sarawak where, moreover the property is situated. The operation of local custom other than English law is limited first to customs of races indigenous to Sarawak, and secondly, to those personal and family relations such as marriage and inheritance of which by English law the test is the domicile of the individual. ... The result is that English law must apply to the facts in the present case.[26]

The subordination of Chinese customary law has to be understood within the context of wider political and economic changes in Sarawak itself, such as the classification of the Chinese as non-native, the Foochow land troubles, the increasing numbers of Chinese settling permanently in the country, the waning power of the *kapitan* and the *kanchu*, and the growth of bureaucracy. With the presence of the Chinese in Sarawak being a *fait accompli*, and with the economic importance and indispensability of the Chinese being emphasized less over the years, Chinese interests could be pushed into the background by the Brooke administration, and Chinese customary law discarded as the body of rules by which the Chinese community was to be governed.

Brooke Reliance upon the Chinese as a Source of Revenue

The first Rajah, James Brooke, had written confidently of the economic role of the Chinese:

Where a Chinaman is found, there the land flourishes, mines are dug, and produce of every description is procured. The Chinese are highly calculated to develop the Dayak, neither have any prejudice of religion, they intermarry, and the Dayaks fall into the category of the Chinese and imitates their industry.[27]

Revenue collected from the Chinese came from several sources, but the main contributions to the Brooke treasury came from direct taxation on the farm monopolies, and import and export duties on goods and commodities traded and planted by the Chinese. The respective contributions of these sources of revenue will be treated in a general way, and it will not be possible to strictly differentiate between what emanated from Kuching as compared with the rest of Sarawak. However, the economic role of the Chinese outside

Kuching, which gradually increased over time, will be emphasized as part of the overall picture.

The early Brooke economy in Sarawak's interior was almost self-sustaining, and as there were no big plantations or mines (except the gold mines in the hands of the Bau *kongsi*) generating revenue or commodities for duty, James Brooke depended heavily upon the monopolies given out for the opium, gambling, pawn and arrack farms which were patronized by the Chinese as a prime source of revenue. In 1859, the farms brought in a revenue of £7,976, against a total revenue of £14,076.[28] The farms were the single most important generator of revenue for the Brookes, and continued to be so until the turn of the century. Table 10.1, albeit incomplete, shows the contribution of the revenue from the farms to the Brooke treasury.

In the nineteenth century, most of the Chinese population was located in Kuching and the surrounding areas of the First Division, where there were concentrations of planters in the outskirts of the capital, in Lundu, and in the gold mines of Bau. The Chinese, through their patronage of the opium, gambling, pawn, and arrack farms, carried the main burden of generating revenue for the Brooke coffers. The monopolies, single or combined, were let out by tender, first for two years, and later for three years. In the case of opium, for example, the successful tenderer then had the right to distribute and sell the opium throughout Sarawak. The farms, in a sense, were a form of indirect tax on the ordinary Chinese *sinkheh* when they bought the opium dross. The monopolies were a lucrative source of revenue for the Rajah, as well as being a profitable business to the individuals and companies which held the rights of sale. Between 1867 and 1910, the monopolies for the farms were held by either leading Hokkien or Teochiu trading firms in Kuching.[29] These companies were also involved in pioneering the cash-cropping of gambier and pepper in the First Division, and they were the sponsors and credit suppliers of rural shops all over Sarawak.

By 1910, the government, convinced that the 'farm' contractors were making too much profit[30] and wishing to obtain more revenue than the tendering system allowed, formed a syndicate partially financed by the Rajah and partially by the local capitalists. It was called the Sarawak Opium Farm Syndicate, with a capital of $150,000, of which the government paid up $60,000, with the remainder borne by Chinese merchants in Kuching. A fixed rent of $24,000 was to be paid monthly to the government, and any profits or losses incurred by the syndicate were to be shared between the government and the Chinese shareholders.[31] This shared venture proved profitable to

TABLE 10.1
Farms Revenue for Selected Years between 1873 and 1896

Year	Revenue by Type of Farm ($)				Total Farms Revenue ($)	Total State Revenue ($)
	Opium	Gambling	Arrack	Pawn		
1873					63,184.00	162,774.50
1874					64,488.30	187,156.98
1876	46,300.00	13,338.00	5,694.47	860.00	66,192.47	170,811.55
1877	54,000.00	14,154.44	5,667.21	1,122.50	74,944.15	185,552.48
1883					127,605.92	271,117.96
1891					211,111.56	456,128.00
1896					205,674.74	573,121.41

Sources: SG, 27 May 1878, p. 35; 22 March 1879, p. 20; 1 October 1883, p. 87; 1 March 1884, p. 20; 2 June 1884, p. 50; and 1 April 1902, p. 62.

the Rajah and, in 1911, a half-yearly rent of $144,000 was paid to the government.[32] The government also held the rights to lease out the farms in the outstations. The syndicate arrangement was renewed for another three years in 1914, with more capital and larger dividends of $34,000 per month to the Brooke administration.[33] The syndicate continued to function until the mid-1920s.

Although the economy was diversifying, it did not mean that the government was prepared to consider the revenue from the monopolies to be of any lesser importance. In 1924, the government took over the Sarawak Opium Farm Syndicate, ostensibly in conformity with policies laid down by the League of Nations, but in reality to have more direct control over this source of income, and to crack down on illicit opium trading.

Immediately after the Brookes assumed control of the monopolies, they were at first able to obtain a substantial revenue, matching that of revenue from import and export duties. But over the years—for example, from 1930 to 1939—the percentage of revenue from the monopolies dropped. This could perhaps have been due to unforeseen administrative costs incurred by the government which, since 1924, had controlled the importation, preparation, distribution, and sale of opium. Table 10.2 below shows the relative revenues earned from the monopolies and from customs duties in selected years. The

TABLE 10.2
Farms and Customs Revenue, 1925–1939

Year	Revenue by Source		
	Farms ($)	Import Duties ($)	Export Duties ($)
1925	1,307,488	442,552	959,460
1930	494,314	946,091	323,863
1935	508,724	1,498,027	250,157
1939	580,977	1,838,253	374,002

Sources: *Sarawak Annual Report of the Treasury for 1925* (Kuching: Government Printing Office, 1926), p. 3; *Sarawak Annual Report of the Government Monopolies for 1930* (Kuching: Government Printing Office, 1931), p. 3; *Department of Trade and Customs and Shipping Office Annual Report for 1930* (Kuching: Government Printing Office, 1931), p. 18; *Annual Report Concerning Prepared Opium for 1935* (Kuching: Government Printing Office, 1936), p. 1; *Sarawak Administration Report for 1935* (Kuching: Government Printing Office, 1936), p. 12; *Annual Report by the Sarawak Government Concerning Prepared Opium in the Sarawak Territory for the Calendar Year 1939* (Kuching: Government Printing Office, 1940), p. 1; and *Supplement to the SGG*, 15 June 1940, p. 34.

TABLE 10.3
Import and Export Duties, 1876–1886

Year	Import Duties ($)	Export Duties ($)	Total Duties ($)	Total Revenue ($)
1876	14,037.81	12,410.80	26,448.61	170,811.55
1877	13,037.91	20,007.95	33,045.86	185,552.48
1878	14,434.31	17,424.01	31,858.32	197,855.52
1882	17,336.56	28,680.40	46,016.96	226,039.55
1883	19,786.49	29,788.72	49,575.21	271,117.96
1885	19,344.39	39,249.38	58,593.77	276,269.18
1886	22,082.83	26,909.52	48,992.35	346,661.71

Sources: *SG*, 27 September 1877, p. 73; 27 May 1878, p. 35; 22 March 1879, p. 20; 1 March 1884, p. 20; 1 May 1886, p. 46; and 1 June 1887, p. 104.

decreasing contribution of the monopolies to Brooke revenue was matched by an increase in customs duties, and outstation Chinese played a major role, contributing with their planting and trading activities.

In retrospect, even in the second half of the nineteenth century, the share of customs duties in Brooke revenue was slowly rising, as Table 10.3 illustrates. These customs duties were generated by the cash-cropping of gambier and pepper, taken up mostly by Chinese cultivators, and by the trade in sago, jungle products, and imported goods. Of course, the Chinese alone were not responsible for the revenue in customs duties. In the expansion of trade, the indigenes—for example, the Melanaus and Ibans—through their enterprise and

TABLE 10.4
Comparisons between Customs and Farms Revenues, 1900–1903

	Revenue by Source		
Year	Customs Duties ($)	Farms ($)	Total Revenue ($)
1900	290,701.11	282,535.00	915,966.00
1901	364,694.65	286,236.80	n.a.
1902	405,770.42	343,657.10	1,192,039.38
1903	482,536.92	398,718.60	1,391,612.68

Sources: *SG*, 1 April 1902, p. 60; and 3 May 1904, p. 100.
Note: Customs duties and revenue from farms do not add up to the total revenue, as income was also derived from other sources.
n.a. = not available

TABLE 10.5
Import and Export Duties, 1910–1920

Year	Import Duties ($)	Export Duties ($)	Total Duties ($)	Total Revenue ($)
1910	151,515.49	351,985.81	503,501.30	1,407,359.58
1911	147,770.96	278,235.66	426,006.62	1,420,420.26
1912	148,040.86	354,306.90	502,347.76	1,462,032.60
1913	138,051.76	130,364.06	268,415.82	1,462,032.60
1914	270,176.72	222,130.25	492,306.97	1,410,797.44
1915	123,595.89	303,554.63	427,150.52	1,536,762.78
1916	124,189.54	291,424.14	415,613.68	1,624,143.71
1917	112,257.62	281,521.74	393,779.36	1,705,292.00
1918	108,342.72	702,355.64	810,698.36	2,531,239.66
1920	186,661.80	878,598.95	1,065,260.75	–

Sources: SGG, 1 May 1911, p. 87; 17 May 1915, p. 128; 16 June 1917, p. 134; 2 June 1919, p. 126; and 1 June 1920, p. 109.
Note: Data for 1919 is not available.

hard work, procured sago from the coast and miscellaneous jungle products from the interior to sell to the Chinese. Many Ibans took up cash-cropping as well. By the beginning of the twentieth century, customs duties had superseded the financial contribution of the farms, which is evident from Table 10.4.

The extension of Sarawak's frontier, which increased trading opportunities, and the opening up of pepper gardens in the First, Second, and Third Divisions were responsible for the big jump in revenue from customs duties. The taking up of rubber planting by Foochow agriculturists and other Chinese and indigenes in the 1910s and 1920s kept the share of customs duties in the total revenue fairly constant. Table 10.5 illustrates the relative importance of customs duties, especially export duties, to the state treasury during that period. It has not been possible to specify in exact terms the overall contribution of the Chinese outside Kuching to Brooke revenue, although certain general indicators have been given. In the first fifty years of Brooke rule in Sarawak, farms—mainly patronized by Chinese from Kuching and the First Division—were a major source of income for the Raj. By the turn of the century and right up to the end of Brooke rule, customs duties from imports and exports, with outstation Chinese playing a big role in planting and trading, were a crucial source of income.

THE CHINESE AND THE BROOKES 217

1. Gertrude L. Jacob, *The Raja of Sarawak: An Account of Sir James Brooke, K.C.B., LL.D., Given Chiefly through Letters and Journals*, Vol. 1 (London: Macmillan, 1876), p. 272.
2. Charles Brooke in 1883 requested the English Consul in Swatow, China, to assist Law Kian Huat to procure coolies while Law was there on a visit. Letters of Rajah Charles Brooke, 10 August 1883, p. 64, SA.
3. See Craig Lockard, 'The Southeast Asian Town in Historical Perspective: A Social History of Kuching, Malaysia, 1820–1970' (Ph.D. thesis, University of Wisconsin, 1973), pp. 155–8; and Craig Lockard, 'Leadership and Power within the Chinese Community of Sarawak: A Historical Survey', *JSEAS*, Vol. 2, No. 2 (September 1971), pp. 195–217.
4. *SGG*, 16 June 1911, p. 110.
5. *SGG*, 16 August 1920, p. 181.
6. Entry dated 17 July 1911 in H.H. The Rajah Confidential, Vol. 9, 1911, pp. 28–9, SA.
7. *SG*, 3 January 1876, p. 108.
8. Entry dated 4 March 1886 in The Rajah's Order Book, Vol. 1, p. 340, SA.
9. *SG*, 3 February 1907, pp. 32–3.
10. 30 September 1896, Baram CCB, p. 73, SA.
11. *SG*, 1 May 1913, pp. 92–3.
12. Letters of Rajah Charles Brooke, Vol. 2, 12 March 1891, p. 175, SA.
13. Letters of Rajah Charles Brooke, 22 March 1896, SA.
14. Lau Ka Too was appointed Kanchu in 1909, at the age of thirty. Unlike most Foochow pioneers, he became a boat hawker and eventually opened a shop in Sibu. As headman of his community, one of his functions was to sponsor new arrivals. For this purpose he had a big house at Sungei Pinjai, used as a temporary lodging- and clearing-house for newcomers. Information on Lau was obtained from his son, Lau Pan Kwong, 20 May 1981, Sibu.
15. *SG*, 2 May 1927, pp. 122–3.
16. *SG*, 1 September 1930, p. 228.
17. Henry Keppel, *The Expedition to Borneo of H.M.S. Dido for the Suppression of Piracy: With Extracts from the Journal of James Brooke Esq.*, Vol. 1 (London: Chapman and Hall, 1846), p. 267.
18. See Chapter 1, section on the *Kongsi*–Brooke Rivalry.
19. T. Stirling Boyd, 'The Law and Constitution of Sarawak', *Journal of Comparative Legislation and International Law*, Vol. 18, Part 1 (February 1936), p. 61.
20. Sarawak Government Orders, 24 January 1881, pp. 324–5.
21. Letters of Rajah Charles Brooke, Vol. 6, 8 July 1910, p. 213, SA.
22. See *SG*, 4 May 1906, p. 126; and 1 December 1916, p. 261.
23. *SG*, 2 November 1922, p. 269.
24. Cited by G. V. C. Young, 'The Transitional Law Period in Sarawak, 1927–1941', *SG*, 30 November 1964, p. 283.
25. *Kho Leng Guan v. Kho Eng Guan*, Supreme Court Records, 1928–1941 (Kuching: Government Printing Office, 1955), p. 60.
26. Ibid., pp. 62–3.
27. John C. Templer (ed.), *The Private Letters of Sir James Brooke, K.C.B., Rajah of Sarawak, Narrating the Events of His Life from 1838 to the Present Time*, Vol. 1 (London: Bentley, 1853), pp. 101–2.
28. FO 12/27, Letter from Spenser St. John to Lord John Russell, 9 August 1860, p. 142.

29. Lockard, 'The Southeast Asian Town in Historical Perspective', p. 201.
30. In 1895, it was estimated that the monopoly farmers made a profit of $50,000. *SG*, 1 May 1896, p. 86.
31. *SG*, 16 June 1910, pp. 129–30.
32. *SG*, 1 May 1911, p. 78.
33. *SG*, 1 May 1914, p. 104.

Conclusion

THE subject of this book has been the rural history of the Chinese in Sarawak during the period of Brooke rule from 1841 to 1941. Hitherto, there has been a historical emphasis in other studies on the mechanics of immigration, and on the Chinese in Kuching and Sibu. This study has set out to provide a historiographical balance to what has already been written about the Chinese experience in Sarawak. Included in the study are groups previously ignored: the *kongsi* miners, the *ulu* traders, and the coolies. The experiences of the Foochow agriculturists have also been reinterpreted against the wider background of social change affecting Sarawak during the earlier half of the twentieth century.

This work has emphasized the pioneering lives of the rural Chinese, and the social and economic adjustments they had to make to their new Bornean environment, beginning with the mining *kongsi* in Bau, proceeding to rural trade and its antecedents, cash-cropping and, finally, the coolies labouring in state-owned and private enterprises. In the pioneers' adjustment to Borneo, relations with the natives were important, and hence this book is also concerned with inter-ethnic relations. On another level, the relationship between the Chinese and a quasi-colonial regime is examined. It is necessary to place the experiences of the Chinese in Sarawak in a comparative context, to highlight similarities and differences where possible with other Chinese communities elsewhere in the *Nanyang*. This will allow for a better appreciation of the history of Chinese settlement in Sarawak.

The Hakka miners in the Bau district came from a West Bornean milieu where the *kongsi* had, since the late eighteenth century, asserted their autonomy against nominal Malay overlords and, later, the Dutch. In Bau, the miners established a *kongsi*, the 'Twelve Company', which was more than just a partnership of interests or a co-operative venture as has been assumed by other observers and authors. It was an instrument of self-government with its own clearly defined territorial jurisdiction, a leadership structure, its own judicial system, its own armaments, and its own currency. Introducing a mining technology of water-gates, sluices and dams, the 'Twelve Company' physically transformed a Land Dayak environment into a

Chinese mining district. The *kongsi* coexisted with its Land Dayak neighbours through intermarriage with Land Dayak women. A mixed community evolved over time, essentially Chinese in cultural orientation and having social and economic bonds with *kongsi* on the other side of the West Bornean watershed. The Bau *kongsi* was a self-sufficient community, and when James Brooke established himself as Rajah in down-river Kuching in 1841, the geopolitical character of the Sarawak River basin was altered.

The 'Twelve Company', in many respects, resembled the *kongsi* on the other side of the Kalimantan frontier. *Kongsi* in Bau and Kalimantan had two distinguishing economic and political features. As Wang Tai Peng has pointed out in his work,[1] the West Bornean (Kalimantan) *kongsi* were rooted in the practices of partnership and brotherhood government (*hui*), and different *kongsi* combined together to form federations. There was a political element to the *kongsi*, beyond the economic goal of pooling resources and sharing profits. *Kongsi* federations in the first half of the nineteenth century—the Ho-shun in Montrado (which later split into the Ta-kang and San-tiao-kou *kongsi*), and the Lan-fang in Mandor—were engaged in a struggle to retain their autonomy against the Malay suzerains and the Dutch, as well as against each other. Just as the Ta-kang *kongsi* and, later, the Lan-fang *kongsi* were to succumb to Dutch pressure, the Bau 'Twelve Company' (an off-shoot of the San-tiao-kou *kongsi*) found it necessary to defend its political autonomy against James Brooke—unsuccessfully, as it turned out.

With the defeat of the Bau *kongsi* in the *Kongsi*–Brooke war of 1857, its effective independence was ended. After 1857, mining continued on a reduced scale, and the district turned to the cash-cropping of pepper and gambier. However, the district gradually came under the direct economic control of the Borneo Company and Hokkien and Teochiu merchants in Kuching. The *kongsi* was a resilient frontier organization which served the needs of the mining pioneers until it was destroyed by Brooke interests, supported by native allies.

As James Brooke expanded his political frontier, other groups of Chinese dispersed throughout Sarawak. The presence of Chinese traders in Sarawak was not a nineteenth-century phenomenon. Archaeological evidence suggests that the Chinese could have traded in North-west Borneo as early as the Tang (AD 618–907) and Sung (AD 860–1279) dynasties. The roots of the present-day Chinese trading presence in Sarawak date from the mid-nineteenth century.

Trade was the speciality of the Hokkiens and Teochius who had a proclivity towards commerce. Bazaars in rural, riverine Sarawak spread through the encouragement and the financial backing of the Kuching Chinese merchants. The extensive riverine trading network centred on Kuching as the entrepôt and hub of commerce. Far more important, however, than the role of the Kuching Chinese from the standpoint of *ulu* trade was the pioneering initiative of the rural traders. Hokkien and Teochiu *sinkheh* were often indebted to their sponsors or, if they were not indebted, preferred to rely upon established or quasi-kin networks for assistance. The greenhorns invariably worked as shop apprentices, or else were asked to open up new shops or become boat hawkers. The knowledge that exotic and assorted jungle products were obtainable in the hinterlands of the Lupar, Rejang, and Baram Rivers was an incentive for the *sinkheh* to travel up-river and become traders. There was a demand for Chinese ceramics and earthenware, luxury goods, and basic foodstuffs from the Ibans and other natives. The hope of being able to eke out a livelihood at the very least, a hope which had brought the newcomers to Sarawak in the first place, compensated for the risks of trading amongst the greatly feared, head-hunting Ibans. Thus, the many small river-front bazaars, together with the ubiquitous boat traders, sprang up in the Lupar, Rejang, and Baram Rivers as a result of Chinese pioneering initiative and the response of the indigenes to the forces of trade.

The relationship of the Hokkien and Teochiu traders with the native people was based on accommodation, albeit an uneasy accommodation. The traders, few in number, lived amongst the native people in what was often a frightening environment. Trading was a dangerous occupation, for traders could literally lose their heads at the hands of the local populace. Hence there was a need for goodwill, tact, and patience, and a willingness to present gifts and forgo debts. The Chinese certainly had to show respect for their clients, for their survival could depend on a show of goodwill and etiquette. They also found it expedient not to flaunt any belongings they had, for it could invite robbery or assault. And, because strength and bravery were qualities admired by the Ibans, the traders found it opportune to demonstrate physical feats of courage and strength whenever the occasion arose.

Despite the acts of violence committed against the itinerant boat hawker, there is no evidence of ethnic group conflict in Sarawak between the Chinese trading minority and the non-Muslim natives. Social theorists would attribute this to the lack of economic competition.[2] It was true that in Sarawak, there was no competition between traders and natives as both occupied and exploited different economic niches. The economic competition theory seems appropriate in explaining conflict between Chinese and Malay traders. The Chinese displacement of Malay commerce, a displacement officially sanctioned by the Brookes, spawned animosity amongst the Malays in outstations like Simanggang, Debak, and Marudi.

The expansion of rural trade in Sarawak paralleled the commercial penetration by the Chinese of provincial areas in other parts of South-East Asia, such as occurred in the Philippines and Thailand. In the middle of the nineteenth century, when the Spanish lifted restrictions on place of residence and commerce for the Chinese in and around Manila, there was a movement into the rural areas. In 1849, 92 per cent of the Chinese population lived in and around Manila; in 1894, only 48 per cent resided in the capital, as the cultivation of abaca (hemp), sugar-cane, and tobacco attracted them to the provinces.[3] When the Bowring Treaty of 1855 opened Siam to world trade, Teochiu and Hakka traders moved up-country from Bangkok, turning small towns into thriving commercial centres.[4]

Regional studies of the Philippines reveal that Chinese traders had successfully established themselves as the buyers of abaca, sugar, tobacco, *tripang* (*bêche-de-mer*), bird's nest, and miscellaneous marine and jungle products, and as the importers and sellers of luxury goods and foodstuffs from the mid-nineteenth century onwards. Jolo in the Sulu Archipelago was one of the most important provincial centres outside Manila; the Chinese from Singapore came to play a crucial economic role in Jolo due to the gradual loss by the Taosug of their trade monopoly as a result of systematic Spanish destruction of Sulu shipping in the late 1870s. Kin and commercial links with Singapore, the setting up of Chinese warehouses in Jolo, and the active collecting, in *prahu*, of marine and jungle products from different parts of the Archipelago contributed to Chinese control of the Sulu Archipelago trade.[5] By the end of the nineteenth century, Chinese traders dominated the local trade in abaca in Bikol,[6] Luzon's south-eastern extreme, and in Samar, the third-largest island in the Philippines;[7] controlled the sugar trade in Pampanga,[8] central Luzon; and monopolized tobacco exports in Cagayan, northern Luzon.[9]

The rapid rise of Chinese dominance in regional trade in South-East Asia was due to certain common factors. In the case of Sulu, it was due to increased Chinese immigration, the operation of an international trading network stretching beyond the Philippines, efficient trading methods in the collection of local products, and an effective retailing system reaching out to the Filipino peasant through the *sari-sari* store. Similar factors were at play in the internal dynamics of *ulu* trade in Sarawak, where the increased tempo of Chinese immigration, the efficient collection of local jungle products from the local people, the selling of imported goods and foodstuffs to the natives, and the role of the ubiquitous boat hawker contributed greatly to a strong Chinese presence in *ulu* trade.

Besides the internal mechanics of rural trade, the successful pioneering trader in *ulu* Sarawak was likely to base his economic relations upon established social bonds. He was likely to be acculturated to the extent of having a local wife, learning to speak the local languages, and possibly adopting local dress and eating local food. Nevertheless, even though the traders were much influenced by the local way of life, they maintained—and were encouraged to retain—a Chinese ethnic identity because official constraints prohibited the development of mixed communities. No 'Sino-Dayaks' were officially recognized; one had either to be a Chinese or a native, and for a Chinese to become a 'Dayak' was forbidden.

Although the traders in Sarawak and their mixed offspring were forced to maintain their Chinese ethnicity and live in bazaars or boats, it would appear from oral traditions and court records that they did not adhere strictly to the rigid regulations of Brooke policies and edicts unless caught. Chinese adaptation to an Iban way of life was conditioned by economic considerations, marital obligations and, in some cases, a genuine attachment to longhouse life. The situation in rural Sarawak was unlike that in the Philippines and Java in the early nineteenth century where *mestizo* and *peranakan* communities, the products of the marital unions of Chinese men and native women, had emerged, acculturated to the local native communities. While no such communities emerged in Sarawak similar to the *mestizo* and *peranakan* because of official disapproval, a large *totok* group, culturally Chinese in every way, was to settle in the Lower Rejang by the turn of the century.

The settlement of Foochow agriculturists in the Lower Rejang was unique. The growth of the Foochow colony was partly due to Brooke policy, and partly because of Foochow initiative. The twentieth-century demand for rubber for the motor-car industry evoked an enthusiasm for planting rubber among the Foochows. This led to economic prosperity for the Foochows and turned them into a closed, self-sufficient community. They built up a network of villages, churches, and schools on their own initiative. Planting rubber trees required extensive areas of land, and it was this craving for land which brought the Foochows into direct conflict with the Ibans. Most writers have pointed out that the different cultural attitudes to land were responsible for the deterioration in ethnic relations in the Lower Rejang.

The Ibans claimed the right to ownership of land for as long as they chose, if they remained in the area, by being the first to clear the primary vegetation. Land was not cleared for permanent cultivation, but would be worked on for a few years and then allowed to lie fallow. To the Foochows, permanent occupation or tenancy was the rule, and land which had lain fallow or unoccupied was considered virgin, unclaimed land.

Conflict over land was of course not peculiar to Sarawak, but was experienced on many pioneering frontiers where there were new settlers, as in Davao on the Philippine island of Mindanao, at the beginning of the twentieth century. Japanese settlers, first brought into Davao to work as road labourers, turned to abaca planting. Their attempts at converting tribal land into abaca plantations brought them into conflict with the Bagobo natives. The Japanese planters were unaware of Bagobo customary land law, with its particular concept of ownership of land and of landmarks such as fruit trees. The American colonial government worsened the situation by classifying areas of communally held land in Bagobo hands as public land, opened to applications from interested parties. The Bagobos, who did not understand the colonial land laws, were unhappy with the Japanese settlers occupying their tribal land, and attacked many of the planters. The 'land problem' which arose in Davao parallels closely the friction over land in the Lower Rejang.[10]

The situations in Davao and the Lower Rejang were basically the same—a conflict of ideas and values between new settlers and natives, with the newcomers not understanding local customs regarding land-ownership and land use. Different cultural attitudes towards land constituted only one factor responsible for ethnic tension in the Lower Rejang. There were others, foremost amongst which was competition for land itself which had become a scarce commodity, not only as a result of the large-scale intrusion of the Foochows, but also because large numbers of Ibans were beginning to plant rubber trees too. The rivalry to open up land, which both Foochows and Ibans needed, accentuated their cultural differences. The theory propounded by researchers like The Siaw Giap[11] of ethnic economic competition leading to communal tension (although The was dealing with trading minorities) would appear to hold for the Lower Rejang. However, there is also support for Mackie's[12] view that ethnic differences and separateness were a 'predisposing' factor in communal relations as both Foochows and Ibans exhibited contrasting cultural attitudes to land.

Another reason for the poor record of Foochow–Iban relations in the Lower Rejang was the nature and scale of Foochow migration. There were parallels to this situation throughout the South-East Asian region by the beginning of the twentieth century. The similarity lay in the mass migration of Chinese to South-East Asia, involving far larger numbers than previously, as well as the migration of Chinese women. Both factors worked against assimilation of the Chinese into the local communities of their host countries.

In the case of Bangkok, Skinner[13] observed that, after 1910, resistance to assimilation on the part of the Chinese increased, as did Chinese awareness of themselves as a separate community. Disregarding changing Thai attitudes and policies towards the Chinese and the influence of political developments in China, the new and ever larger waves of migrants, including more women, contributed to the Chinese becoming a separate economic and social community. This was the case, too, with the Lower Rejang Foochows who came as an almost complete community, with their wives, children, kin, and quasi-kin. There was little need for the Foochows to enter into relationships with their neighbours, and they were not dependent on the trader–Iban bond which involved a continuous selling–buying relationship accentuated by social ties.

The educational and social structure of the Foochows encouraged them to retain their 'Chineseness'. Lockard has shown that mission schools in Kuching, through the medium of English language, had an integrative role in bringing pupils of different races together, and these bonds continued after school, in working life, and in social and sporting activities.[14] This does not seem to have been the case in Sibu, where the Methodist schools, situated right in the heart of the Foochow villages and using both Chinese and English as mediums of instruction, were the sole preserve of Foochow students until after the Second World War, when the Methodist secondary schools in Sibu town itself began to attract pupils of other races. Church, school, and community were linked in a tightly spun web, woven around itself at the expense of other sub-ethnic and ethnic groups. The presence of associations which catered to common commercial interests, county origins, and *alma mater* ties encouraged adherence to Chinese cultural values and social networks limited to kin and members of one's own dialect group.

The social and cultural resistance of the Foochows in Sarawak can be compared with information from two case-studies on Java. Willmott's *The Chinese of Semarang*[15] and Tan Giok Lan's *Chinese of Sukabumi*[16] point to the social and cultural differences, resulting from the larger number of migrants within a short period, between the acculturated, *peranakan* and the new *totok* communities. The *totok* were the 'pure' or 'full-blooded' Chinese, whose arrival at the beginning of the century, together with Chinese women, enabled the continuation of an 'unadulterated' Chinese society and culture. Unlike the earlier batches of Chinese men who took native wives, the *totok* did not find it necessary to do so. The children of the *totok* were brought up to speak Chinese and imbued with Chinese philosophical and religious ideas, and they rejected *peranakan* culture which was considered 'adulterated'. Chinese education in Mandarin was provided by the *Tiong Hoa Hwe Koan* (Chinese Association); previously, only rudiments of education had been provided in dialect. The upsurge of interest in Chinese education was indicated by the number of Chinese schools—seventy-five altogether with some 5,500 pupils in 1908.[17] Not surprisingly, in the Lower Rejang, the Methodist schools maintained a strong interest in Chinese education and served as a strong transmitter of Chinese social and cultural values, which made the Foochows even more insular.

The reaction of the Brookes to the Foochows, their separateness, and their land problems with the Ibans, was to impose immigration cuts and restrictions on rubber production, and introduce legislation regarding land-ownership and nationality. The effect of all these measures was to make the Foochows more culturally tenacious as a Chinese sub-ethnic group.

While the Foochows represented a closed, agrarian-based community, other Chinese enclaves had been established in other parts of Sarawak through another form of pioneering activity—wage-labouring in the coal mines and oilfields. The Brookes had experimented with coal mining in the Sadong in the mid-nineteenth century and had relied upon indentured Chinese labour. When the oilfields in Miri were opened at the beginning of the twentieth century, Chinese workers were sought to perform the hazardous semi-skilled and skilled jobs.

Among the various types of ways of earning a living that were available to the Chinese pioneers in Sarawak, wage-labouring was one of the toughest, due to the indentured form of recruitment and employment which kept coolies in conditions of near servitude at low wages for indefinite periods of time. Working and living conditions were harsh and dangerous, resulting in industrial accidents, illness, deaths, and suicides. There was little social life for these workers who were virtually all bachelors and for whom gambling and opium-smoking were probably the only forms of social recreation available. The labourers fought for higher wages and better working conditions, and to free themselves from bondage, but did not succeed. This was due in large measure to the Brookes being able to rely on the voluntary services of the Ibans who, being naturally inclined to warlike activity, helped to clamp down on industrial strife.

The employment of coal mine and oilfield labourers in Sarawak can be compared to the employment of Chinese labour in the gold mines of the Transvaal, South Africa, during the period 1904 to 1910. The Transvaal coolies were brought in as indentured labourers, under conditions similar to those which applied to coolies brought into Sarawak. Working and social conditions in the Transvaal were similar to those experienced by the Sadong and Miri labourers. In the Transvaal, there was a high fatality rate from accidents and suicides, and injuries and illnesses were commonplace. It had been estimated that out of a peak labour force of 15,427 in 1906, as much as one-quarter of the work-force was convicted for various misdemeanours ranging from absence without leave to refusal to work.[18] The reasons for such a high rate of offences lay in low wages, a poor working environment, and excesses committed by the management in supervising the workers. Similar conditions in Sarawak were behind the labour troubles in Sadong and Miri. While the Sarawak labourers turned to gambling and opium-smoking, the Chinese in Transvaal had male prostitution[19] in addition to gambling as social outlets. The experiment with Chinese labour in the Transvaal lasted only six years; complaints in the British Parliament and the ready availability of an African work-force helped to terminate an unjust indentured labour system involving imported Chinese labourers. No such forces were at work in Sarawak: for a start, the Sarawak Raj was a private dynasty, not a Crown Colony, and the local people could not be enticed in sufficient numbers to replace the Chinese labourers.

Thus far, an account has been given of the diverse pioneering experiences of the Chinese outside Kuching, ranging from mining to trading, planting and labouring. The spread of the Chinese into rural Sarawak was a dynamic process, conditioned by the physical environment, economic forces, and the demands of the Brookes for outside labour. The social and economic history of the rural Chinese has been placed within the wider framework of Sarawak society and history. The history of the rural Chinese in Sarawak is not a separate history in itself. There was considerable interaction and tension between the Chinese and the indigenous groups inhabiting Sarawak's rivers and jungle, and between the Chinese and the Brookes. Elements of accommodation and conflict emerged; on balance, however, harmony and coexistence seem to have prevailed, and there appears to have been more harmony, for which credit is to be given to the attitudes of the Chinese pioneers and the native people themselves.

The major contributions of the Chinese pioneers outside the secure base of Kuching were: first, the economic role they played in contributing sorely needed revenue to the Brookes through their multiple economic skills and through their patronage of the opium and other farms; and second and more importantly, the role they played in building up this state, by putting down roots and opening up new frontiers all across Sarawak.

A high percentage of Sarawak's ethnic Chinese population still live in the rural areas of the state, working as farmers, traders, and labourers. There continues to be a great deal of intermarriage between the Chinese and the natives, and many Chinese can speak the local languages. The Chinese continue the erstwhile pattern of accommodation with the local people. There is an absence of conflict where there is no competition for resources between the Chinese farmer or trader and the native. Where the needs and interests of the Chinese do not encroach upon those of the indigenes, mutual tolerance develops. Even in the towns with their high concentrations of Chinese residents, education and exposure to others at work and play have encouraged ethnic tolerance. This is most evident during the celebration of major festivals like Chinese New Year, Gawai, and Hari Raya, when the different races visit each other at home.

Where there is competition over resources or when there is a perceived need to protect a community's wider interests to the exclusion of others, then co-operation and social relations between the different races may be threatened. This is evident at the formalized level of political activity—that is, at the level of political parties.

During Sarawak's early post-war period (that is, in the 1940s and 1950s), multiracial co-operation existed within political parties like the Sarawak United People's Party (SUPP), formed in June 1959. Although the party was Chinese-controlled with strong Hokkien and Hakka support, it nevertheless had substantial native membership and support. It became identified as a Chinese party rather than one that also represented native interests partly as a result of pressure applied by the colonial government to erode native support for the party, and partly after the party formed a coalition government in 1970 with Parti Bumiputera when it began to fight more outspokenly for Chinese interests. (The leadership of SUPP had decided that it could better fight for Chinese interests by aligning itself with the Malay or Muslim-led Parti Bumiputera.[20])

Another example of successful multiracial co-operation within a political party is the Sarawak National Party (SNAP) established in April 1961 by a group of Ibans from the Saribas district of the Sri Aman Division of Sarawak, all employed by Brunei Shell Company. After 1966, when Chief Minister Dato Stephen Kalong Ningkan from the Iban-controlled party was expelled from the state government for espousing the cause of Sarawak nationalism, the party gathered considerable Chinese and Muslim support. The party garnered even more Chinese support when, after 1970, it took up the cause of Sarawak nationalism more vigorously as an opposition party to the coalition government formed between the SUPP and Parti Bumiputera.[21] As with SUPP, the strains in multiracial co-operation began to emerge, in this case over the issues of Dayak leadership of the party and Dayak rights.[22] The Sarawak National Party lost a number of Iban supporters in 1983 when dissident Iban leaders and members decided to form the communal Parti Bangsa Dayak Sarawak.

Within Chinese society itself, the identification of occupation with dialect group no longer holds due to the impact of education and the availability of new economic opportunities. However, the rural element is still strong and many Hakkas and Foochows continue to be smallholders, and many shopkeepers in the *ulu* are still Teochius and Hokkiens. A major structural and social change affecting the Chinese has been the aggressive inroads made by the Foochows, beyond their rubber smallholding background, into the areas of trade, banking, real estate, timber logging, and other sectors of the modern economy. But this is not to imply that the other dialect groups have not also moved into other occupations and areas of economic activities: they have, though on a far lesser scale and at a lesser intensity.

The rapidity of Foochow occupational diversification and, with it, their control of certain economic activities, like timber logging and banking, are reminiscent of their early history in the Lower Rejang. In the first half of the twentieth century in the Lower Rejang, the Foochows' requirement for land for permanent cultivation, their feelings of insecurity, and their desire to constantly expand and protect their economic niche without due regard for native land rights brought them into conflict with the Ibans. In recent years, for example in 1987, conflict has again occurred over land-related issues—this time, the demands of the timber-logging industry versus indigenous land rights.[23] The present conflict is more complicated than previous conflicts in that the present timber concessions are held not only by the Chinese but also by élite groups from other communities.

Still, this does not negate the point that it is instructive to understand the present timber-logging disputes in Sarawak by examining the earlier land disputes in the Lower Rejang at the turn of the century. It may be revealing to note that some present-day attitudes towards land use may not differ from attitudes held earlier in the century. For example, it is still believed that shifting cultivation as practised by the natives is 'wasteful' as compared to the 'usage of land on a more permanent basis' by the Chinese.[24]

When the collective experiences of the Chinese are assembled and reviewed, their economic and social contributions to the development and present make-up of Sarawak cannot be ignored. The past contributions of the Chinese miners, traders, planters, and labourers, especially the ordinary *sinkheh* who played such an important pioneering role, need to be remembered and put alongside those of the other communities, all of whom make Sarawak what it is today.

1. Wang Tai Peng, 'The Chinese Republic in West Borneo from the Latter Part of the 18th Century to the Middle of the 19th Century' (MA thesis, Australian National University, 1977).

2. See W. F. Wertheim, *East–West Parallels: Sociological Approaches to Modern Asia* (The Hague: W. van Hoeve, 1964).

3. Edgar Wickberg, *The Chinese in Philippine Life, 1850–1898* (New Haven: Yale University Press, 1965), pp. 61–2.

4. G. W. Skinner, *Chinese Society in Thailand, An Analytical History* (Ithaca, New York: Cornell University Press, 1957), pp. 86–8.

5. James Warren, *The Sulu Zone, 1768–1898: The Dynamics of External Trade, Slavery and Ethnicity in the Transformation of a Southeast Asian Maritime State* (Singapore: Singapore University Press, 1981), pp. 5–17, 126–48.

6. Norman Owen, 'Kabikolan in the Nineteenth Century: Socio-economic Change in the Provincial Philippines' (Ph.D. thesis, University of Michigan, 1976), pp. 117–19.

7. Bruce Cruickshank, 'Continuity and Change in the Economic and Administrative History of 19th Century Samar', in Alfred McCoy and Ed. C. de Jesus, *Philippine Social History: Global Trade and Local Transformations* (Quezon City: Ateneo de Manila University Press, 1981), p. 238.

8. John Larkin, *The Pampangans, Colonial History in a Philippine Province* (Berkeley: University of California Press, 1972), pp. 48–53.

CONCLUSION 231

9. Ed. C. de Jesus, 'Control and Compromise in the Cagayan Valley', in McCoy and de Jesus, op. cit., p. 31.
10. Cecil Cody, 'The Consolidation of the Japanese in Davao', *Comment*, Vol. 7, No. 3 (1958), pp. 26–36; and Shinzo Hayase, 'Tribes, Settlers, and Administrators on a Frontier: Economic Development and Social Change in Davao, Southeastern Mindanao, The Philippines, 1899–1941' (Ph.D. thesis, Murdoch University, 1984).
11. The Siaw Giap, 'The Chinese in Indonesia', *Kabar Seberang*, Vol. 7 (1980), pp. 114–30.
12. J. A. C. Mackie, *The Chinese in Indonesia* (Melbourne: Angus and Robertson, 1974), p. 129.
13. Skinner, *Chinese Society in Thailand*.
14. Craig Lockard, 'The Southeast Asian Town in Historical Perspective: A Social History of Kuching, Malaysia, 1820–1970' (Ph.D. thesis, University of Wisconsin, 1973), pp. 295–312.
15. D. E. Willmott, *The Chinese of Semarang: A Changing Minority Community in Indonesia* (Ithaca, New York: Cornell University Press, 1960).
16. Tan Giok Lan, *The Chinese of Sukabumi: A Study of Social and Cultural Accommodation*, Modern Indonesia Project (Ithaca, New York: Cornell University Press, 1963).
17. Mary F. Somers Heidhues, *Southeast Asia's Chinese Minorities* (Melbourne: Longman, 1974), p. 39.
18. Chen Ta, *Chinese Migrations, with Special Reference to Labour Conditions* (Washington, DC: Government Printing Office, 1923), p. 131.
19. Persia Crawford Campbell, *Chinese Coolie Emigration to Countries within the British Empire*, 2nd edition (London: Frank Cass, 1971), p. 211.
20. See Michael B. Leigh, *The Rising Moon, Political Change in Sarawak* (Sydney: Sydney University Press, 1974), pp. 8–22, 142–61. Parti Bumiputera merged with Pesaka, an Iban party, in 1973 to form Pesaka–Bumiputera Bersatu.
21. Ibid., pp. 32–5, 102–12.
22. *Sarawak Tribune*, 15 September 1983.
23. See *BB*, 27 June 1987; 11 July 1987; 1 August 1987; and 8 August 1987; and *Asiaweek*, 28 June 1987.
24. Letter from Major (Ret.) K. K. Ngu, in *BB*, 29 August 1987. This letter was written in response to remarks by Datuk Amar James Wong, Environment and Tourism Minister in the Sarawak State Government, on the practice of shifting cultivation by Ibans in Limbang, near his (Datu Amar Wong's) timber logging concession.

Glossary

Abang	hereditary aristocratic title
adat	native customary law
attap	roofing thatch
Awang	traditional title of aristocrats
bandong	Chinese sailing junk, boat
batang	river
baju	shirt
batu	stone
belian	ironwood, a species of timber hardwood
bejalai	journeying for the purpose of procuring jungle products or to find work
bieng-dang	pole slung over shoulders for carrying goods or parcels
bilek	room in a longhouse
blachu	black cotton cloth
bukit	hill
bunkal	a unit of weight for gold
chanang	gong
chap ji ki	lottery game
chawat	loincloth
chelapa	oblong box for keeping accessories
chita	cotton chintz
chue-tsai	'piglet', term of reference for indentured labour
damar	jungle resin
damun	(Iban) secondary jungle
datching	scale for weights
Datu	respectful title for a man of high status or rank
Datu Bandar	'Port Chief'—the most powerful Malay leader in Brooke Sarawak
Datu Hakim	'Judge'
Dayang Muda	wife of the Rajah's heir (*see* Tuan Muda)
depas	measure of an arm's length
engkabang	species of tree local to Borneo yielding fruit with high fat content; the fruit is used in soap and candles

GLOSSARY

fan tan	game of chance in which cards are used
fu-t'ou-jen wei-ko	assistant headman
gantang	unit of weight or volume
Gawai	Iban rice harvest festival now celebrated as the Iban new year
gelong	strips (rattan)
gutta baru	adulterated jungle gum
gutta percha	jungle resin
Hari Raya	a festival, especially the celebration at the end of the fasting month for Muslims
hsien	county
hui	brotherhood government
huo-chang	overseer
hua-chiao	overseas Chinese
illipenut	species of tree local to Borneo yielding fruit with high fat content; the fruit is used in soap and candles (the plant is locally also known as *engkabang*)
jala	barges
jangkar gutta	a type of jungle resin
jelutong	jungle resin
kain asap	nankeen cloth
kajang	mat protection, for example, for roofs
kampong	(Iban) primary vegetation
kampong	(Malay) village
kapitan	Chinese headman
kanchu	river chief; also, a Chinese headman
kepala China	decapitated head of a Chinese
kongsi	form of open government based on enlarged partnership and brotherhood
kuala	river-mouth
kubu	fort
Kwan Yin	Goddess of Mercy, one of the popular goddesses in the Chinese pantheon
labong	forge iron
laksa	vermicelli dish served with vegetables and spicy, chilli-hot gravy
landas	wet season from October to March
langkau	hut

GLOSSARY

lo-tai	elder brother
loteng	loft
mandor	foreman
masok Melayu	conversion to Islam
minyak tanah	oil
mestizo	person of mixed Chinese–Filipino descent, acculturated to Filipino native society
nakoda	captain or master of boat
nanga	mouth of stream or river
Nanyang	literally 'South Seas'; term of reference for South-East Asia
negri	country
padi	rice before threshing, in the husk
Pangawa	district head
pangkalan	river landing place
Penembahan	'Object of Veneration'; a Malay title
parang	long, sheathed knife
parit	mine or canal
passu	unit of measurement
penghulu	Iban headman
peranakan	(Indonesian) local-born Chinese, speaking the Indonesian language, sometimes of mixed descent, and partly acculturated to native society
pikul	a unit of weight
prahu	boat
Rajah	ruler
Rajah Muda	Heir Apparent
rattan gemaing	jungle creeper
rattan sega	jungle creeper
rempah	spices and condiments
ruai	communal veranda in longhouse
sampan	boat
samseng	gangster
samshu	rice wine
sari-sari	(Filipino) miscellaneous goods store
sarong	Malay garment comprising strip of cloth tucked round waist or under armpits
shan-sha	'mountain of sand'
sessar	dried prawns
sinkheh	literally 'new guest'; new arrival from China

GLOSSARY

sireh	betel-leaf for chewing
sungei tanah	river document
surat mudai	document
surat tanah	land grant
sut	outer clothing
ta-ko	eldest brother
Tai Pek Kong	
(Tua Pek Kong)	Chinese deity deifying the spirit of pioneers
tang	hill
tanjong	headland, promontory
tapang	tall Bornean timber tree
ta-tsuen-chieh	presidential system or republic
ta-tsung-chang	great president
tawak	gong
'tebassed'	cleared
tian	house support
Tien Ti Hueh	(Chinese) triad society
totok	(Indonesian) 'pure' Chinese, foreign-born, Chinese-speaking, and less acculturated
towkay	Chinese shopkeeper
tripang	bêche-de-mer
tua kampong	native headman
tuai rumah	headman of a longhouse
Tuan Muda	(lit.) Young Lord
tukang ayer	water carrier
Tu Ti	(Chinese) 'Earth God'
ts'ai-ku	clerk
ulu	hinterland, or up-river, up-country
undang-undang	Malay law code
wayang	open air theatre

Appendices

APPENDIX 1

List of Interviewees

AGES of interviewees are given as at 1981, when the interviews were conducted.

A. Kuching

Liew Peck Kwee

Hakka, and a retired schoolteacher with a keen interest in Sarawak history, he is author of *The Hoppo Chinese, with Special Reference to Sarawak* (Singapore: Tung Yet Publishing House, 1978).

B. Simanggang

Goh Khai Huat

Teochiu, 75 years old, and born in Simanggang, he had been a shopkeeper. Goh was interviewed in an old men's home, living out his last days.

Lim Gu Piao

Teochiu, 86 years of age, Lim first arrived in Spaoh in 1920 and became a boat hawker. He was a good example of a *sinkheh* forced to remain behind, poverty-stricken, and a victim of gambling losses.

Tan Khee Kiang

Hakka, and 82 years of age, Tan was born in Sabu, a few miles outside Simanggang. Sabu was a Hakka planting settlement.

Tay Bak Choon

Teochiu, 93 years of age, this *sinkheh* reached Simanggang in 1910. He was a remarkable informant with a fine memory of his days as a boat hawker.

Tay Ing Boon

Teochiu, 81 years old, the informant had lived in Simanggang since 1921, and had worked as a boat trader.

APPENDICES   237

Madam Tay Lan Eng

Teochiu, 72 years of age, and born in Lubok Antu, she still had clear memories of Lubok Antu when she was a young girl. Madam Tay moved to Simanggang in 1926.

Yeo Yiaw Piaw

Teochiu, 71 years old, Yeo arrived in Simanggang in 1927. He was a boat hawker.

C. Engkilili

Bong Nam Siong

Hakka, 72 years of age, and born in Engkilili, Bong was a reliable source of oral recollections on the Hakka mining company in Marup, near Simanggang—the 'Fifteen Company'—and on trade in Engkilili.

Tok Sia Hwang

Teochiu, 64 years of age, Tok was a *sinkheh* who came to Engkilili in 1928 to join his father. Tok became a boat hawker.

D. Lubok Antu

Lim Thai King

Ninety-four years old, Lim was a third-generation Hakka in Sarawak. Lim could speak fluent Iban. When interviewed, he could still relate anecdotes about his life as a trader, as well as stories of how trade was carried out in Lubok Antu.

E. Betong

Lim Chen Tau

Hokkien, and 83 years old, Lim had been living in Betong since 1924. He had been a shopkeeper.

Sim Ah Lah

Hokkien, 87 years old, Sim was born in Kuching, but moved to Betong in 1914. He had been a trader.

Tan Tze Liong

Teochiu, 63 years of age, he came to Betong in 1927 to join his father. A dentist by profession, Tan was a knowledgeable oral source on the history of Betong.

F. Roban

Ng Seng Phua

Hokkien, 73 years old, Ng arrived in Roban as a young child in 1913, brought over by his father. He was a pepper planter before he took up boat trading.

G. Debak

Ling Yang King

Teochiu, 78 years of age, Ling landed in Debak in 1926 and worked as a shop apprentice for some kinsmen.

H. Saratok

Ong Tiang Boon

A Hokkien-speaking Foochow, 79 years old, with a reliable memory, Ong was a greenhorn when he journeyed to Saratok in 1916. He made a living as a boat hawker.

I. Sibu

Kho Cheng Kang

Hokkien, 77 years of age, Kho was brought to Sibu from Fukien when he was still a young child. He worked as a shop apprentice in Singapore, Kapit, and Sibu before finally opening his own shop in Sibu in 1935.

Kong Chung Siew

Cantonese, 62 years old, Kong was born in Sibu. He became a community leader of the Cantonese and was a knowledgeable source of local history of the Cantonese settlement in Sibu.

Lau Pan Kwong

Foochow, 65 years old, Lau was born in Sibu. His father was the well-known *kanchu*, Lau Ka Too.

Lau Tzy Cheng

Foochow, a local amateur historian, and a prolific writer, Lau has had many articles published in local Chinese newspapers as well as several books to his credit.

Law Hieng Ing

Foochow, 81 years old, Law was a schoolteacher in Foochow before being recruited in 1926 to teach in Sibu. His life-long interest in community affairs made him a reliable source of local Sibu history.

Ling Chu King

Foochow, 72 years of age, Ling arrived in Sibu as a *sinkheh* in 1927. Ling initially worked as a wage labourer for a Cantonese rubber planter. In 1936, in partnership with three others, he started a boat service from Sibu to Sarikei, Binatang, and Kapit.

Siaw Tiang Choon

A locally-born Hokkien, 72 years old, Siaw's grandfather came to Sibu in 1876 by jumping ship. Siaw was a boat hawker.

Tan Tong Chee

Teochiu, 84 years of age, Tan arrived in Sibu in 1930 to set up his own business.

Madam Law Suok Moi

Foochow, 58 years old, Madam Law arrived in Sibu in 1940 and worked as a rubber tapper for a relative.

Tay It Leong

Hokkien, 33 years old, Tay represented the fourth generation of his family in Sarawak. His great-grandfather had been a trader in Kanowit.

Ting Ming Hoe

Foochow, 76 years old, Ting was born in Sibu. Ting's grandfather was among the first batch of Foochow pioneers.

Wong Kwong Yu

Foochow, 64 years of age, Wong was taken to Sibu in 1926 as a young child by an uncle. He first worked in his uncle's rubber garden, but subsequently became a schoolteacher.

Wong Neng Sung

Foochow, 74 years old, Wong went to Sibu in 1938 for an unusual reason—to look for a debtor who owed him money.

J. Binatang

Sim Ngiik King

Foochow, 77 years old, Sim went to Binatang in 1928 to take up rubber planting.

Teng Tun Hsin

Foochow, 75 years of age, Teng was born in Sibu. Teng's parents arrived with the second batch of Foochow pioneers. He attended a local school at Sungei Sadit, Sibu. In 1929, he joined the Land Office as a clerk, and resigned in 1939 to start his own business.

K. Sarikei

Chieu Pak Ming

A Cantonese community leader, 53 years old, Chieu represented the second generation of his family in Sarawak. His father had arrived in Sarikei in 1917. Chieu was well informed on the local history of the Cantonese in Sarikei.

Hii Hua Siong

Foochow, 76 years of age, Hii arrived in Sarikei in 1925, starting off as a shop apprentice, then becoming a schoolteacher, and later a headmaster. Hii displayed a keen interest in local history.

Wong Ngiong Hua

Foochow, 78 years old, Wong migrated to Sibu in 1912 as a young child. His father was one of the Foochow pioneers of Sarikei.

L. Kanowit

Lim Choon Jin

Hokkien, 71 years old, Lim represented the second generation of his family in Sarawak. Lim's grand-uncle sojourned to Kanowit in 1879, followed by his father in 1888.

Lim Zean Khuan

Hokkien, 51 years of age, Lim represented the fourth generation of his family in Sarawak. His great-grandfather arrived in Kanowit in 1880. Lim himself only set foot in Sarawak in 1949, but was nevertheless well informed on the history of Chinese settlement in Kanowit.

Siaw Kee Siong

A locally born Hokkien, 81 years old, Siaw had made a living as a boat hawker.

Tan Cheng Swee

A Hokkien *kapitan*, 73 years of age, Tan was born in Sibu and only moved to Kanowit in 1936.

M. Kapit

Syn Chin Joo

A second-generation Hokkien, 65 years of age, Syn had considerable knowledge of Kapit's past.

Teo Tien Siong

A second-generation Teochiu, 64 years old, Teo had been a boat hawker in his younger days.

N. Belaga

Neo Teck Hua

A second-generation Hokkien, 72 years of age, Neo had been a boat hawker. He had a good memory for his own family history and the local history of Belaga.

O. Miri

Chai Shun Fatt

A Sabah-born Hakka, 62 years of age, Chai attended a mission school in Kuching, and then joined the Anglo-Saxon Petroleum Company as a clerk. He displayed a reliable memory for early social life in Miri.

Chan Wey Yuen

Shanghainese, 80 years of age, Chan arrived in Miri in 1930 and worked as an artisan. He displayed a good knowledge of social conditions in Miri before the Second World War.

Chew Peng Man

Fifty-seven years of age, Chew was the son of a Cantonese *mandor*, Chew Fong.

Chu Nyin Hee

Cantonese, 69 years old, Chu arrived in Miri in 1922 as a young boy to join his father who had been there since 1917. Chu's father had been a partner in a restaurant business, and Chu himself was working as a goldsmith when interviewed.

Kong Pit Leong

Foochow, 78 years old, Kong first started working in Miri in 1935, at the age of 32, and was employed as a cook for the company medical officer of the Anglo-Saxon Petroleum Company.

Madam Ku Nam Siong

Hakka, 81 years of age, Madam Ku was born in Kuching in 1900, but moved to Miri at the age of 21 to marry a trader.

Lau Hong Onn

Hokkien, 70 years of age, Lau had been in Miri since 1925, when he started working life as a shop apprentice.

Lee Yok Seng

Cantonese, born in Kuala Lumpur, Lee arrived in Miri in 1933, and found employment as a clerk in the Anglo-Saxon Petroleum Company.

Reverend John Leong

Kuching-born Hakka, 56 years old, living in Miri since 1933, Reverend Leong was an invaluable source of information on pre-war Miri.

Leong Kam Fook

Cantonese, 77 years of age, Leong went to Miri in 1927 at the age of 23. He was employed as an electrician by the Anglo-Saxon Petroleum Company.

P. Marudi

Chang Teng Seng

Hokkien, 80 years of age, Chang had been a boat hawker during his youth. Chang's father had been a partner in the well-known Marudi shop of Joo Watt.

APPENDICES

Ho Hong Kwang

Locally born Teochiu, 82 years old, Ho received a mission-school education in Marudi. He joined the Anglo-Saxon Petroleum Company in Miri in 1927 as an apprentice draughtsman.

Lim Poh Kui

Locally born Hokkien, 65 years of age, Lim was educated at a mission school in Kuching. Coming from a prominent family of Marudi traders, he turned to trading in 1936.

Tien Chu Kim

Locally born Teochiu, 76 years old, Tien turned to boat hawking in 1919, at the age of 14, travelling between Marudi and Long Lama.

Q. Long Lama

Kang Chiat Sam

Hokkien, 76 years of age, Kang arrived in Marudi in 1934, sponsored by a Long Lama trader, Lee Chin Tek, for whom he subsequently worked. Kang was well informed on Long Lama's local history.

Yap Poh Chai

Hokkien, 62 years old, Yap was born in Marudi. His father was a trader in Long Akah, and Yap himself became a boat hawker in 1936, at the age of 16.

APPENDIX 2

Trade on the Kalaka and Saribas Rivers, 1907

THIS account of the bazaars in the Kalaka (Krian) and Saribas river basins, written in 1907 by D. J. S. Bailey, Resident in Simanggang, reveals much about the structure of trade stretching all the way from the rural hinterland to Kuching, the kind of goods traded, the competition between the Chinese and Malays, and the physical conditions of the shophouses.

THE KALAKA (KRIAN)

1. The Kabong Bazaar consists of eight shops fairly well built of *belian* wood and roofed with *belian* shingles. As it has not increased in size for many years, one might presume that the trade there remains much the same, but, as a matter of fact, it has of late undergone some marked changes. A few years ago the Kabong Chinese bought and salted fish for export to Kuching; now it does not pay them to do so and the fish curing is in the hands of the Malays who sell their fish inland, and up and down the coast, little going to Kuching unless the Kuching Malays go to Kabong to buy it. The Kabong Chinese, however, still purchase the dried prawns (*sessar*) on the coast, chiefly at Sungei Samaludam, and export them in great quantities to Kuching. There is also the *jelutong* and *jangkar gutta* trade which has increased a good deal of late and every Kabong Chinese trader goes in for it.
2. The next bazaar to notice in this river is that at Saratok, which, I am sorry to say, still remains the same cluster of insanitary leaf-built shophouses which the Chinese have inhabited since the (second) fire of September 8th, 1901 destroyed their *belian*-built bazaar (which was the finest building of the kind in the Kalaka river).

    The Saratok trade appears to be much as it was with the exception that sago is not worked freely. Still, the working of the common *guttas* is more lucrative and, until these are worked out, the sago export will continue to be depressed. There are a few pepper plantations at the back of the station, but they are not large nor very flourishing. In addition to the export of *gutta* and sago, during good years, there is always a very fair amount of padi and rice taken out from Saratok, and there should be a good export of these in 1907 if the harvest is got in well.
3. There remains the Roban bazaar. It consists of eighteen shophouses of which thirteen are fairly well built of *belian*. The trade done at Roban is much the same as in the Saratok bazaar with the exception that there is a quantity of black and some white pepper from the gardens in the Ibus and near Roban, and an uncertain amount of *belian* sawn-planks and scantlings.

    There are six Chinese *bandongs* running from the Kalaka to Kuching (three from Saratok, and three from Roban) and they seem to be fairly regular and to take good cargoes. In addition to these are Malay-owned *prahu* of about five tons burthen, the exact number of which varies a great deal, but of late they have certainly increased in number. The *nakodas* of these boats take *gutta* to Kuching and bring back, rice, *kainkain*, oil, tobacco, etc.; which their owners peddle for *gutta*.

THE SARIBAS

1. Curiously enough there is the same number of Chinese-owned *bandongs* in the Saribas river as there is in the Kalaka, three of these at Betong, and three at Pusa. There are also some large Chinese-owned 1 masted boats and several Malay-owned 5 ton *prahu*.

   The trade at both Betong and Pusa is much the same as it has been in its good years for some time past. There are practically no Chinese engaged in pepper planting in the Saribas—indeed there is not much room for them in this river.

   Of course the trade in the Saribas has always been inferior to the Kalaka as the river is smaller and there are fewer people.

   The bazaar at Pusa is inferior to the Betong bazaar and always has been so, even if the small Debak bazaar be included with that of Pusa.

   I do not see that there is likely to be any great expansion of trade in these two rivers unless some new industry of minerals is found. In good padi years, we shall have good exports of rice and padi, and if there is no sickness or disturbance, we may expect that the exports and imports will exceed those of the very best years we have enjoyed—by some small amounts—but nothing more.

   The welfare of the people, taken as a whole, and their standard of living, have undoubtedly improved since I have known these rivers. Further improvements, which must necessarily be very slow, depend entirely upon the advance the people themselves make in farming and planting. This advance is retarded chiefly by the communal system of land tenure, and the Dayak custom of living in 'long' houses (which paralyses all individual initiative). With these obstacles removed, or modified in some way, we might expect a satisfactory advance in agriculture.

Source: SG, 5 June 1907, p. 138.

APPENDIX 3

Boat Cargo, 1922

A boat owned by a Sibu shopkeeper, sailing from Singapore to the Natuna islands, was accidentally blown off course to Mukah. It fortuitously carried a full cargo list which is reproduced here. There is little reason to believe that the types of goods it carried would have been significantly different from that taken by other boats.

Rice 26 bags	$288.02
Tobacco 60 bundles	27.00
Flour 2 bags	9.70
Onions 20 katis	3.06
Sessar (dried prawns) 10 katis	8.80
Cakes 700 bundles	9.45
Fish 10 katis	3.70
Tea 2 boxes	5.30
Rempah 71 katis	7.45
Needles	2.50
Rempah 10 katis	3.00
Buttons 1 packet	.50
Brooms	1.40
Pencils	1.20
Fish hooks 48 bundles	1.85
Fishing line 26 bundles	7.20
Cloth 3 *kayu*	10.50
Food 6 katis	39.50
Cigarette paper 10 bundles	1.80
Soap 10 packets	7.00
Laksa 25 katis	4.00
Onions 37 katis	4.64
Cakes 500	13.00
Kerosene oil 10 tins	65.40
Roti (biscuits) 50 katis	10.50
Cotton 72 reels	8.10
Books 24	1.20
Enamelled ware	3.60
Condensed milk	22.50
Soap 2 boxes	3.60
Cloth 10 *kayu*	31.50
Kechup (soya sauce) 60 bottles	6.80
Laksa 10 katis	1.55
Brasswire	2.81
Empty boxes	1.50
Cash Dutch money	337.50
Cash Straits money	362.50

Source: 'Teo Chong Loh for Chop Soon Bee v. Chan Boh', 20 March 1922, Sibu CCB, p. 304.

APPENDIX 4

The Burning of Sibu Bazaar, 1889 and 1928

ULU bazaar fires were a common occurrence. The way in which shops were constructed, the building materials used, the storage of inflammable goods in the shops, and carelessness in the kitchens at the back of shophouses all contributed to the frequency of fires. The following account describes the outbreak of two fires in Sibu.

FIRE AT SIBU, 1889

Another disastrous fire has again occurred in Sarawak, whereby the whole bazaar at Sibu was destroyed in a few hours. After Kuching, the bazaar at Sibu is, or rather was, the finest and most busy in the country carrying on almost the whole of the trade of the Rejang river. Piles of *rattans* and other jungle produce are always lying stacked there, awaiting shipment and the shops are well stocked with trade goods. The immediate cause of the fire is not certain, but it is reported to have originated in an opium shop, that fruitful cause of trouble; later reports say in a kitchen; this is equally likely as the style of kitchen range in vogue amongst Sarawak Chinese is of wood covered with a coating of clay or cement. The Sibu bazaar was built of *belian* planks only, but these and the workmanship employed to put them up were of the best quality and the majority of the shops were comparatively new. Such buildings, when once alight, burn rapidly and so fast did the flames spread in this case that but little was saved. Though the outbreak of fires cannot be stopped by legislation, the risk might be diminished in some degree by the issuing of stringent regulations with regard to the placing of lamps, storing of dangerous goods, and the use of suitable buildings. Thus if all opium smoking shops were bound to be brick buildings or even corrugated iron, and cooking stoves were only placed at a safe distance from inflammable walls, many fires which have but small beginnings, but which increase in size owing to the ready supply of good wood for the flames, could be extinguished before much damage had been done. The method of building brick houses also leaves much to be desired, whole streets being sometimes erected without, in the upper storey, a single brick partition to separate house from house and even where there are such brick partitions, they reach only to below the tiles, and the beams of the roofs invariably overlap each other in a manner which is certain to allow and promote the spread of fire from one house to its neighbour, and so on for the whole length of a street.

We do not know whether fire insurance companies care to take such risks as these, but be that as it may, there is no insurance effected on Chinese property here. This should at least cause care to be taken by owners and occupiers, but precaution against danger from fire appears to be entirely wanting amongst the Chinese, who however in the presence of an actual outbreak are too panic-stricken and excited to do anything to help check the flames. The want of water at a high pressure renders the extinction of fires by this method extremely doubtful and the two large fires which have previously occurred here, the burning of the Kuching bazaar in 1884, and the sago factories in 1888, were only prevented from doing much greater damage, by sheer luck; in the former case by a heavy rain storm coming down in the nick of time and in the latter by the wind being in the right direction; had it veered round, the whole range of factories would have been destroyed on the one side, or the Malay *kampong* of Padungan on the other. In the case of Sibu no such good luck intervened; consequently a clean sweep was made of the whole place and many have been, at least for the present, entirely ruined; it will be years before they can make good their losses, if they ever do so at all, by legitimate trading, and many of those who less than a month ago called themselves *towkays* will now have to start afresh from the level of coolie.

FIRE AT SIBU, 1928

On the morning of the 11th ultimo at 11.30 a.m. a fire broke out in the up-river end of the Sibu bazaar, and in the space of about half an hour the whole of that end of the bazaar was in flames. As the wind was blowing up-river it was hoped that the fire could be checked, but as usual in such cases, the Chinese were soon in despair, and confined their attention to saving goods. With the aid of Malays and Dayaks, several attempts were made to check the fire by pulling down some of the houses, but these attempts were of no avail, owing to the wind shifting, and the intense heat and the difficulty experienced in dismantling these strongly built houses, and by 6 a.m. the fire had burnt out, having completely destroyed the bazaar. The fire, as far as can be ascertained to date, arose from carelessness in not putting out a kitchen fire after use that evening. In some cases the inmates had no time to save anything, in most cases very little was saved, and even the greater part of what was brought out of the houses and piled on the road caught fire and was destroyed. The police station was also destroyed. Sibu bazaar consisting of some 60 houses, was recently rebuilt and was entirely made of *belian* wood which, though slow to ignite, when once on fire burns fiercely. The loss in house property, trade goods and personal effects amounts to about $150,000, a severe loss to the unfortunate traders of this district, and it will be some time before a similar bazaar is erected. The Sibu traders hope to be able to build the new bazaar entirely of brick.

Source: SG, 2 April 1928, p. 67.

APPENDIX 5

Rejang Settlement Order, 1880

It was in 1880 that Charles Brooke first announced his plans to attract the Chinese to the Rejang as planters.

I, CHARLES BROOKE, Rajah, make known the following terms which the Government of Sarawak hereby agrees to fulfil with any Company of Chinese who will engage to bring into the Rejang River Chinese settlers with wives and families numbering not less than three hundred souls, who will employ themselves in gardening and farming padi or in other cultivations—

1st—The Government will provide land sufficient for their requirements free of charge.

2nd—The Government on first starting will build them temporary houses, and make a good path to their landing place.

3rd—The Government will give them one *passu* of rice per man or woman a month and a little salt and half the amount to every child for the first 12 months.

4th—The Government engage to keep up steam communication with Kuching and carry any necessaries for these settlers on the most reasonable terms.

5th—The Government will build a Police Station near them to protect them and assist in making themselves understood in the native language and generally look after them.

6th—In carrying out the above engagements the Government expect the said Chinese will permanently settle in the territory of Sarawak.

Kuching, 11 November, 1880.

Source: *SG*, 29 November 1880, p. 59.

APPENDIX 6

Agreement between Charles Brooke and Wong Nai Siong for the Immigration of Foochows, 1900

MEMO OF AGREEMENT was made in duplicate between the Sarawak Government on the first part, and Messrs. Nai Siong and Tek Chiong of Chop Sim Hock Chew Kang, hereinafter mentioned as the Contractors on the other part.

1. The Contractors agree to introduce into the Rejang River 1,000 adult Chinese agriculturists, men, women, and about 300 children and to establish them in that river for the purpose of cultivating rice, vegetables, fruits, etc., but of these immigrants not more than one half are to be introduced during the first year, that is to say before June 30, 1901, and the rest the contractors undertake to introduce during the following year, that is to say between June 30, 1901 and June 30, 1902.
2. The Government undertakes to advance the Contractors the sum of $30 for each adult and $10 for each child so introduced, and of these advances, two thirds shall be paid to the Contractors in Singapore, and the balance at Kuching on the arrival of the immigrants there, and the Contractors undertake that the majority of the immigrants to be introduced during the first year as mentioned in Paragraph 1, shall be brought to their destination in the Rejang within 4 calendar months from the date they receive the advances in Singapore as mentioned above.
3. The Contractors undertake to repay all such advances to the Government within 6 years from the date of this Agreement as follows: Nothing to be paid by the Contractors during the first year; during each of the subsequent years one-fifth of the advances to be paid each year, that is to say $6 for each adult and $2 for each child in respect to the advances paid on their account in accordance with Section 2.
4. The Government undertakes to provide for the passages of the aforementioned immigrants from Singapore to the Rejang, or, in the event of the Contractors bringing these immigrants direct from China to the Rejang, the Government will pay the Contractors $5 for each immigrant as passage money.
5. The Government undertakes to provide the Contractors free of all rent or other charges for the term of 20 years from the date of this Agreement, sufficient land in the Rejang in the vicinity of Ensurai and Seduan streams, or elsewhere, for the proper settlement of the aforementioned immigrants and to insure that the immigrants shall get sufficient land for their purpose, the quantity of land being not less than 3 acres for each adult.
6. On the expiration of the above-mentioned term of 20 years any immigrant shall on his application be given a grant for the land occupied by him subject to quit rent at the rate of 10 cents per acre per annum, provided that such land be fully cultivated.

7. In the event of the Government wishing to occupy any land taken up by any of the immigrants, a fair sum shall be paid to such immigrants by the Government for disturbance in respect to crops, houses, etc.
8. The Government undertakes to make suitable landing places, roads, and paths.
9. On the recommendation of the Contractors, the Government will recognize the appointment of any competent and suitable man as *kanchu* or headman of each village or settlement. The power of such *kanchu* will be limited to the settlement of trivial disputes, boundary disputes, and other minor matters, but these powers will be more clearly defined by the Government when necessity subsequently arises for their appointments.
10. The Government guarantees full protection to immigrants from interference by Natives.
11. The Government will place no restrictions on the immigrants with respect to their planting or the scale of their produce, and they will be at liberty to plant what they please and sell where they like, but it is understood by the Contractors that the primary object in introducing these immigrants is the cultivation of rice, and they, on their part, undertake to see that this is not lost sight of.
12. The Government undertakes to ship all provisions, stores, etc. for the immigrants and produce sent by them to Kuching, on Government vessels at moderate rates of freight as opportunities of shipping by Government vessels afford, but the Government does not undertake to run steamers especially for the purpose of carrying such goods and produce, but will do its best to assist the immigrants in this respect.
13. The Government will not permit any persons to visit the immigrants for the purpose of inducing them to gamble or to gamble with them, nor to sell opium to them. Gambling amongst the immigrants may be allowed, or not as decided to be advisable by the Government and the Contractors, and, if at any time it is allowed, it will be confined solely to immigrants under the supervision of their headmen, and such headmen will alone have the right to sell opium to the immigrants under their charge. The Government will make special arrangements with the farmers from time to time to insure those rules being carried out effectually.
14. The Government will permit that a limited but sufficient number of muskets may be kept by the immigrants to protect their crops from the ravages of wild pigs, etc.
15. After the expiration of 2 years from the date of this Agreement, should the immigrants be successful and their settlements be in a thriving condition, the Government will permit others joining them from China, and will assist such fresh immigrants in as far as it may lay in its power to do so.
16. Should the Contractors be successful in carrying out the objects in respect to these immigrants, and succeed in establishing prosperous settlement or settlements, they will be permitted to conduct such

trading operations as they may wish, and successful planters will be permitted to trade.

17. On their sureties for the repayment of the advances as agreed upon and mentioned in Paragraph 3 of the Contractor's offer:

Signed—Khoo Siok Wan, Lim Boon Kheng

and the said Khoo Siok Wan and Lim Boon Kheng do hereby affix their seals and sign their names as having duly given security for the Contractors in this respect.

Signed, sealed and delivered on the 9th day of July, 1900. (Signatories are Khoo, Lim, Wong Nai Siong and Tek Chiong, Contractors: Charles Brooke and C. A. Bampfylde, Government: George Muir of Paterson, Simons & Co., Witness; at Kuching, July 27, 1900).

Source: Rajah's Agreement Book, April 1893–December 1902, SA.

APPENDIX 7

Cultivation of Rice in the Rejang

REVEREND HOOVER, the successor of Wong Nai Siong, attempted to explain the Foochows' rice planting methods during their first few years in the Rejang. He then made suggestions on how rice cultivation could be undertaken.

According to an agreement made by the late Rajah with Mr. Wong Nai Siong, a Foochow colony was established on the Rejang in 1901, "for the purpose of cultivating rice and vegetables", (See Agreement, Art. 1). There is an impression abroad that these colonists never made an honest effort to plant rice. This is a mistake.

1. The first year they all planted rice at Seduan on the Sungei Merah, but they used seeds they had brought from China, and being seeds adapted to a colder climate they matured too soon. When the stalks were only a foot high they begin to head, the heads did not mature, and the crop was not worth harvesting.
2. The second year they tried again planting two crops. The first crop was a failure because it was planted too early in the season to make way for the second crop; the second brought nothing principally because of the birds which concentrated on it, being the only rice in the neighbourhood.
3. The third year they tried the Dayak method of just cutting and burning without cultivating (as in former years they had done in China, and expended much time and labour). As might be expected, this failed completely.
4. The fourth year they were so discouraged and homesick that they refused to plant, and lived almost entirely at the expense of the proprietor. This soon brought him to bankruptcy, and the colony to the verge of failure. At this junction, the Rajah thought the colony would do better without the proprietor, and he was removed from the country.
5. The next year it was necessary to plant rice or starve as there was no proprietor to supply it, and, as it happened, they got a good crop.
6. By this time, many had begun to plant pepper, and in a short time, everybody was planting as much as it was possible to take care of.
7. For the next two or three years, the rice crops were more or less of a success, but the time came when it did not pay to husk the rice, a man's labour was worth more doing something else. It took a man and two women a day to husk and clean a bag of rice. At the same price rice was selling in the bazaar this did not pay, and compared with what a man could earn sawing boards or planting potatoes, he was losing money. This difficulty was solved by the introduction of a rice-huller.
8. Rice cultivation then took a new life with some hope of supplying the market of Sarawak. But there were many failures of crops from reasons I [Reverend Hoover] will name later. Very seldom was a full crop harvested.

9. Then came rubber, and the planting of rice gradually fell off, because when people have money as these people have, it is much cheaper to buy rice than to raise it; always providing there is any to buy.
10. Then came the war, and last year, the bazaar ran short of rice several times. With much persuasion and some fear, everybody put out rice, but the crop, with very few exceptions, is a failure. When the rice was about eight inches high, floods came and submerged it for more than a week, leaving if half drowned, from which it did not recover. Now here we are with rice dearer and scarcer here than it has been before, and we have no rice of our own. Malay and Milanau rice here is a total failure.
11. With the years and experience in rice raising I have had in this country, I may be excused if I try to explain how I think rice can be successfully cultivated.

1. It cannot be done the way the natives do it. It is a gamble in which they lose nine times in ten. They never get what might be called a full crop.
2. It does not pay to do it by hand even by the best methods under the best supervision, except of course in starving times like this when rice is not to be had at any price. Mr. Davis of our Mission tried it out at Bukit Lan; a hundred weight of padi cost him $18, when rice was selling at $8 a hundred weight in the bazaar. This of course was a bad year; but it never pays, labour is too dear, and other things pay better, even working *jelutong*.
3. Nevertheless, if rice could be planted with a reasonable hope that a full crop would be harvested, a lot more would be planted.
4. The things that make rice planting so precarious an undertaking are worms, grasshoppers, birds, pigs, deer, floods and drought.
    (a) Worms can be dealt with. I have seen them gathered by the bucketful by holding the bucket under the stalk, and striking the stalk with a stick, the worms falling into the buckets.
    (b) Grasshoppers and other flying or hopping insects may be killed by passing a torch between the rows of growing rice at night. The insects jump into the fire as it passes along and make good fertilizer; I have seen fields infested with them cleared in one or two nights.
    (c) Birds, pigs and deer can be kept off by pickets, poisons, scarecrows, etc. At most, they would not cause much damage to a big field.
    (d) If floods come at the right time and are not too high, they make the crop. If they are early, they prevent burning and planting. If they are too late, ploughing and weeding cannot be done. The only way to get away from the floods is to select land that is never flooded more than one foot at highest tides.
    (e) Irrigation is the only remedy for drought. This could be easily done with pumps, as the water would never have to be lifted more than three or four feet, and a little water would go a long way on such level land.

To sum up, the following are necessary:-
1. Cultivate in large tracts.
2. Select land of proper elevation with regard to floods.
3. Use machinery—tractors, ploughs, etc.
4. Irrigate when necessary.
5. Competent supervision.

Source: *SG*, 2 June 1919, p. 140.

APPENDIX 8

Opening of the New Chinese Court, 1911

THE Chinese Court, established from 1911 until 1920, gave the Chinese jurisdiction over Chinese customs and customary law. An account of the official opening of the new Chinese Court House follows.

The new Chinese Court House which has just been completed was formally opened by His Highness the Rajah on the 1st instant. His Highness, with the *Tuan Muda* and *Dayang Muda*, arrived at the entrance at 10 a.m. The open space in front of the building had been roped off and inside this enclosure were gathered the Chinese Magistrates of the Court, the *Datus*, and nearly all the Europeans resident in Kuching, including several ladies, while the general public assembled outside. His Highness was met at the entrance to the building by Mr. H. F. Carew-Gibson who presented a handsome silver key of local Chinese workmanship with which the Rajah proceeded to open the main door of the new Court House. Everyone followed His Highness into the building and when all were seated, His Highness addressing the company in English, said that for a long time past he had been thinking out some way of improving the status of the Chinese community who were very useful inhabitants of an Eastern country and in whose hands was a large part of the trade of this country. The Government had always been more or less at a disadvantage in deciding cases in which Chinese laws and customs had to be taken into account, cases relating to marriage, divorce, inheritance, etc., and therefore, a year ago, had instituted a Chinese Court, to be presided over by Chinese, for the hearing of such cases, and as a help to the Superior Courts in such matters as they might require advice and assistance. This Court had been in existence for one year during which time it had worked satisfactorily and had been of great assistance to the Government in dealing with purely Chinese affairs. This Court was presided over by Mr. Ong Tiang Swee assisted by six Chinese Magistrates drawn into different clans and elected annually. He [Charles Brooke] had now had a proper Court house built for them which he now declared open. His Highness then made another speech in Malay, after which Mr. Ong Tiang Swee replied suitably on behalf of the Chinese community. This ended the short ceremony.

The new Court House is situated on the bank of the river near the bridge crossing Sungei Kuching, on the spot where the stone crusher used to be. It is a handsome square building of concrete blocks and bricks with two wings. The main hall is about 46 feet square inside, and the two side rooms are 15 feet square each. There are two doors 8 feet wide in front facing the road, and two behind facing the river, with stone steps leading up to them, and in addition there are eight windows opening to the floor with ornamental iron railings, thus ensuring plenty of light and air in the building. It should be mentioned that the silver key presented to the Rajah by the Commissioner was designed in the Public Works Office and executed by Messrs. Kong

Chan & Co., 9 Carpenter Street. The design consisted of an open book bearing the word *"Lex et Justitia"* and the scales of justice, supported by the Sarawak and Chinese Republican flags, the whole surmounted by a badger, the crest of Sarawak.

Source: *SG*, 16 July 1912, p. 156.

Bibliography

Unpublished Primary Records

Great Britain

Rhodes House Library, Oxford

MSS. Pac. s 90, Basil Brooke Papers.
Correspondence and Papers of the Brooke Family of Sarawak including papers of Charles T. C. Grant (1831–91), Laird of Kilgraston c.1830–1977.

Colonial Office

CO 802/1 and CO 802/2. Sessional Papers: Sarawak Annual Reports, 1900–41.

Foreign Office

FO 12. Borneo, 1847–1875.

Borneo Company Limited, London

Martine, T. C., 'History of the Borneo Company Ltd'. From 'Notes written by me when in Singapore (Changi Gaol) 1943–4' (typescript).

Malaysia

Sarawak Museum, State Archives, Kuching

Letters of Rajah Charles Brooke and Rajah Muda Vyner Brooke, 10 vols., 1880–1915.
Simanggang (Second Division) Court Records, 33 vols., 1864–1930.
Second Division, Copy Cash Book, 1927–8.
Second Division, Police Court Book, 1920–3.
Betong (Second Division) Court Records, 12 vols., 1866–1930.
Sibu (Third Division) Court Records, 7 vols., 1874–1923.
Third Division, Court of Requests (debt cases), 4 vols., 1914–22.
Baram (Fourth Division) Criminal Cases, 9 vols., 1882–1918.
Fourth Division, Police Court Books, 1919–23.
Fourth Division, Civil Cases, 1883–1916.
Fourth Division, Probate Books, 1899–1919.
Agreement Book (records of contracts and other agreements between the government and companies and individuals), 4 vols., 1877–1922.
Chinese and Native Employees Roll Book, 1 vol., 1890–1909.

Hammond, R. W., 'Report on Education in Sarawak' (unpublished typescript, 1937).
His Highness The Rajah Confidential (mostly reports from the Resident of the First Division, including commentary on events throughout Sarawak), 2 vols., 1904–6 and 1911–13.
His Highness The Rajah's Order Book, 4 vols., 1863–1920.
Letter Book (reports from A. B. Ward, Resident of the First Division, to the third Rajah on affairs throughout Sarawak), 1918–22.
Mimeographed newsletter marked 'Confidential to Government Officers', Nos. 1–9, January–October 1938.
Roll Book No. 1, Records of European officers on Permanent Service, undated.

Library of Malaysian Rubber Research Institute, Kuala Lumpur

Bridges, W. F. N., 'A Report on Rubber Regulation in Sarawak' (unpublished typescript, 1937).

Published Official Records

Orders and Laws

Orders which have not since been cancelled, issued by H.H. the Rajah or with his sanction from 1863 to 1916, inclusive (Kuching, 1891–1917), 10 vols.
Index to State Orders (Kuching, 1933).
State Orders Enacted during the Years 1936 to 1941, inclusive.
Boyd, T. Stirling (comp.), *The Laws of Sarawak, 1927–1935* (London: Bradbury, Wilkinson, 1936).
Hedges, R. Y. (comp.), *The Laws of Sarawak in Force on the Second Day of July 1947*, 3 vols. (Kuching: Government Printing Office, 1948). Three additional volumes were published from 1949 to 1951.
Richards, A. J. N. (comp.), *Dayak Adat Law in the First Division, Adat Bidayuh* (Kuching: Government Printing Office, 1964).
―――― *Dayak Adat Law in the Second Division* (Kuching: Government Printing Office, 1961).
―――― *Sarawak Land Law and Adat* (Kuching: Government Printing Office, 1961).
The Penal Code, 1934 (Kuching: Government Printing Office, 1934).
Sarawak Government Orders 1860–1890. One volume indexed. Orders and miscellaneous material pertaining to the administration of Charles Brooke, as Tuan Muda, and later as Rajah.
Sarawak Administration Reports, 1927–1938 (Kuching, 1928–39).
Sarawak Civil Service List, 1925 (Kuching, 1925). Also issued in 1926–30, 1932, 1934, and 1937–9.
Supreme Court of Sarawak, Reports of Decisions (Kuching: Government

Printing Office, 1928). Three more volumes were published in 1936, 1939, and 1946.

Rubber Regulation during 1936 (Kuching, 1937). A similar report was issued for 1937.

Reports

Allied Geography Section, Southeast Pacific Section, *Area Study of Sarawak and Brunei*, Vol. 1 (September 1944).

Clark, C. D. Le Gros, *1935 Blue Report* (Kuching: Government Printing Office, 1935).

Department of Statistics, 'State Population Report, Sarawak', in *Population and Housing Census of Malaysia* (Kuala Lumpur: Government Press, 1983).

Jones, L. W., *Sarawak: Report on the Census of Population taken on 15 June 1960* (Kuching: Government Printing Office, 1962).

Noakes, J. L., *Sarawak and Brunei, A Report on the 1947 Population Census* (Kuching: Government Printing Office, 1950).

Porter, A. F., *Land Administration in Sarawak* (Kuching: Government Printing Office, 1967).

Wilford, W. E., *The Geology and Mineral Resources of the Kuching–Lundu Area, West Sarawak, including the Bau Mining District* (Kuching: Government Printing Office, 1955).

Newspapers and Newsletters

Borneo Bulletin
Salam (Sarawak Shell Berhad newsletter)
Sarawak Gazette
Sarawak Government Gazette
Sarawak Tribune

Theses

Austin, Robert F., 'Land Policy and Social Change: The British Presence in Sarawak', MA thesis, University of Michigan, 1974.

────── 'Iban Migration: Patterns of Mobility and Employment in the Twentieth Century', Ph.D. thesis, University of Michigan, 1977.

Chang Tsuen-Kung, 'Historical Geography of Chinese Settlement in the Malay Archipelago', Ph.D. thesis, University of Nebraska, 1954.

Cotter, Conrad, 'Some Aspects of the Administrative Development of Sarawak', MPA thesis, Cornell University, 1955.

Diu Mee Kuok, 'The Diffusion of Foochow Settlement in the Sibu–Binatang Area, Central Sarawak, 1901–1970', MA thesis, University of Hawaii, 1972.

Dixon, Gale, 'Rural Settlement in Sarawak', Ph.D. thesis, University of Oregon, 1973.

Doering, Otto C., III, 'The Institutionalization of Personal Rule in Sarawak', M.Sc. Econ. thesis, London School of Economics, 1965.

Fidler, Richard, 'Kanowit: An Overseas Chinese Community in Borneo', Ph.D. thesis, University of Pennsylvania, 1973.

Fortier, David, 'Culture Change among Chinese Agricultural Settlers in British North Borneo', Ph.D. thesis, Columbia University, 1964.

Gill, Sarah, 'Selected Aspects of Sarawak Art', Ph.D. thesis, Columbia University, 1968.

Hayase, Shinzo, 'Tribes, Settlers and Administrators on a Frontier: Economic Development and Social Change in Davao, Southeastern Mindanao, The Philippines, 1899-1941', Ph.D. thesis, Murdoch University, 1984.

Hensen, Erik, 'Iban Belief and Behaviour', Ph.D. thesis, Oxford University, 1969.

Lockard, Craig, 'Chinese Immigration and Society in Sarawak, 1868-1917', MA thesis, University of Hawaii, 1967.

_____ 'The Southeast Asian Town in Historical Perspective: A Social History of Kuching, Malaysia, 1820-1970', Ph.D. thesis, University of Wisconsin, 1973.

Moench, Richard, 'Economic Relations of the Chinese in the Society Islands', Ph.D. thesis, University of California, 1966.

Omohundro, John Thomas, 'The Chinese Merchant Community of Iloilo City, Philippines', Ph.D. thesis, University of Michigan, 1974.

Owen, Norman, 'Kabikolan in the Nineteenth Century: Socio-economic Change in the Provincial Philippines', Ph.D. thesis, University of Michigan, 1976.

Rousseau, Jerome, 'The Social Organization of the Baluy Kayan', Ph.D. thesis, University of Cambridge, 1974.

Siaw, Laurence, 'A Local History of the Ethnic Chinese Community in Titi, Malaysia: Circa 1870 to 1960', Ph.D. thesis, Monash University, 1975.

Sutlive, Vinson, 'From Longhouse to Pasar: Urbanization in Sarawak, East Malaysia', Ph.D. thesis, University of Pittsburgh, 1972.

Wang Tai Peng, 'The Chinese Republic in West Borneo from the Latter Part of the 18th Century to the Middle of the 19th Century', MA thesis, Australian National University, 1977.

Warren, James, 'Trade, Raid and Slave: The Socio-economic Patterns of the Sulu Zone, 1770-1898', Ph.D. thesis, Australian National University, 1975.

Books

Baring-Gould, S. and Bampfylde, C. A., *A History of Sarawak under Its Two White Rajahs 1839-1908* (London: Sotheran, 1909; reprinted Singapore: Oxford University Press, 1989).

Beeckman, Daniel, *A Voyage to and from the Island of Borneo* (London: Dawsons, 1973; first published 1718).

BIBLIOGRAPHY

Belcher, Edward, *Narrative of the Voyage of H.M.S. Semarang, during the Years 1843–46*, 2 vols. (London: Reeve, Benham and Reeve, 1848).

Boyle, Frederick, *Adventures among the Dayaks of Borneo* (London: Hurst and Blackett, 1865).

Brassey, Thomas, *Voyages and Travels of Lord Brassey, K.C.B., D.C.L., from 1862 to 1894*, edited by Captain S. Eardley-Wilmot, 2 vols. (London: Longmans Green, 1895).

Brooke, Charles A. J., *Queries: Past, Present and Future* (London: The Planet, 1907).

⎯⎯⎯ *Ten Years in Sarawak*, 2 vols. (London: Tinsley, 1866).

Bunyon, Charles John, *Memoirs of Francis Thomas McDougall, sometime Bishop of Labuan and Sarawak, and of Harriette His Wife* (London: Longmans Green, 1889).

Burkill, I. H., *A Dictionary of the Economic Products of the Malay Peninsula*, 2 vols. (London: Crown Agents for Colonies, 1935).

Campbell, Persia Crawford, *Chinese Coolie Emigration to Countries within the British Empire*, 2nd edition (London: Frank Cass, 1971).

Cartwright, Frank T., *Tuan Hoover of Borneo* (New York: The Abingdon Press, 1938).

Cavenagh, Colonel Orfeur, *Report upon the Settlement of Sarawak* (Calcutta: J. Kingham, 1863).

Chau Ju Kua, *Chu-fan-chi*, translated by Friedrich Hirth and W. W. Rockhill (St. Petersburg, 1911).

Chen Ta, *Chinese Migrations, with Special Reference to Labour Conditions* (Washington, DC: Government Printing Office, 1923).

Chin, John, *The Sarawak Chinese* (Kuala Lumpur: Oxford University Press, 1981).

Chin, Lucas, *The Cultural Heritage of Sarawak* (Kuching: Sarawak Museum, 1981).

Clarkson, James D., *The Cultural Ecology of a Chinese Village: Cameron Highlands, Malaysia*, University of Chicago, Department of Geography, Research Paper 114 (Chicago, Illinois: University of Chicago Press, 1968).

Collingwood, Cuthbert, *Rambles of a Naturalist on the Shores and Waters of the China Sea* (London: John Murray, 1868).

Comber, Leon, *Chinese Secret Societies in Malaya: A Survey of the Triad Society from 1800–1900* (New York: J. J. Augustin, 1959).

Crawfurd, John, *A Descriptive Dictionary of the Indian Islands and Adjacent Countries* (London: Bradbury and Evans, 1856).

⎯⎯⎯ *History of the Indian Archipelago*, 3 vols. (Edinburgh, 1820).

Crisswell, Colin, *Rajah Charles Brooke, Monarch of All He Surveyed* (Kuala Lumpur: Oxford University Press, 1978).

H.H. The Dayang Muda of Sarawak (Gladys Milton Brooke), *Relations and Complications* (London: John Lane, 1929).

De Windt, Harry, *My Restless Life* (London: Grant Richards, 1909).

⎯⎯⎯ *On the Equator* (London: Cassel, Peter and Galpin, 1882).

Denison, Noel, *Jottings Made during a Tour amongst the Land Dayaks of Upper Sarawak, Borneo, during the Year 1874* (Singapore: The Mission Press, 1879).

Digby, K. H., *Lawyer in the Wilderness*, Cornell University, Southeast Asia Program Data Paper No. 114 (Ithaca, New York: Cornell University Press, 1980).

Dunn, F. L., *Rain-forest Collectors and Traders: A Study of Resource Utilization in Modern and Ancient Malaya*, JMBRAS, Monograph No. 5 (Kuala Lumpur: 1975).

Earl, George Windsor, *The Eastern Seas: Or Voyages and Adventures in the Indian Archipelago, in 1832-33-34* (London: William Hallen, 1837; reprinted Singapore: Oxford University Press, 1971).

Emerson, Rupert, *Malaysia: A Study in Direct and Indirect Rule* (Reprint edition, Kuala Lumpur: University of Malaya Press, 1964).

Foggo, George, *Adventures of Sir James Brooke, K.C.B., Rajah of Sarawak* (London: Effingham Wilson, 1853).

Forrest, Thomas, *A Voyage to New Guinea and the Moluccas from Balambangan, including an Account of Magindano, Sooloo and Other Islands* (London: G. Scott, 1779).

Freedman, Maurice, *Chinese Family and Marriage in Singapore* (London: HMSO, 1957).

―――― *The Study of Chinese Society* (Stanford: Stanford University Press, 1979).

Freeman, Derek, *Iban Agriculture: A Report on the Shifting Cultivation of Hill Rice by the Iban of Sarawak* (London: HMSO, 1955).

―――― *Report on the Iban of Sarawak* (Kuching: Government Printing Office, 1955).

Furness, W. H., *The Home Life of Borneo Headhunters, Its Festivals and Folklore* (Philadelphia: Lippincott, 1902).

Geddes, W. R., *The Land Dayaks of Sarawak: A Report on a Social Economic Survey of the Land Dayaks of Sarawak*, presented to the Colonial Social Science Research Council (London: HMSO, 1954).

Gomes, Edwin, *The Sea Dayaks of Borneo* (London: Society for the Propagation of the Gospel, 1907).

―――― *Seventeen Years among the Sea Dayaks of Borneo* (London: Seely and Co., 1911).

Griffiths, Sir Percival, *A History of the Inchcape Group* (London: 1977).

Groeneveldt, W. P., *Historical Notes on Indonesia and Malaya, Compiled from Chinese Sources* (Jakarta: C. V. Bhratara, 1960).

Gullick, J. M., *Indigenous Political Systems of Western Malaya* (London: The Athlone Press, 1958).

Haddon, Alfred C., *Headhunters, Black, White and Brown* (London: Methuen, 1901).

Haddon, Alfred C. and Start, Laura, *Iban or Sea Dayak Fabrics and Their Patterns* (Cambridge: Cambridge University Press, 1936).

Harrisson, Tom, *The Malays of Southwest Sarawak before Malaysia: A Socio-ecological Survey* (London: Pall Mall, 1970).
_____ *World Within: A Borneo Story* (London: Crescent, 1959).
_____ (ed.), *The Peoples of Sarawak* (Kuching: Government Printing Office, 1959).
Heidhues, Mary F. Somers, *Southeast Asia's Chinese Minorities* (Melbourne: Longman, 1974).
Helms, Ludvig Verner, *Pioneering in the Far East, and Journeys to California in 1849, and to the White Sea in 1878* (London: W. H. Allen, 1882).
Hooker, M. B., *A Sourcebook of Adat, Chinese Law and the History of the Common Law in the Malayan Peninsula* (Singapore: University of Singapore, 1967).
Hornaday, William, *Two Years in the Jungle* (London: K. Paul, Trench, 1885).
Hose, Charles, *Fifty Years of Romance and Research: Or a Jungle Wallah at Large* (London: Hutchinson, 1927).
_____ *Natural Man: A Record from Borneo* (London: Macmillan, 1926; reprinted Singapore: Oxford University Press, 1988).
_____ *The Field-Book of a Jungle Wallah* (London: H. F. & G. Witherby, 1929; reprinted Singapore: Oxford University Press, 1985).
Hose, Charles and McDougall, William, *The Pagan Tribes of Borneo*, 2 vols. (London: Macmillan, 1912).
Hume, Joseph, *A Letter to the Right Honourable the Earl of Malmesbury, Secretary of State for Foreign Affairs, etc., Relative to the Proceedings of Sir James Brooke, K.C.B. etc., in Borneo* (London, 1853).
Ingleson, John, *Expanding the Empire: James Brooke and the Sarawak Lobby*, (Perth: University of Western Australia, Centre for South and Southeast Asian Studies, 1980).
Ireland, Alleyne, *The Far Eastern Tropics: Studies in the Administration of Tropical Dependencies* (Boston: Houghton Mifflin, 1905).
Irwin, Graham, *Nineteenth Century Borneo: A Study in Diplomatic Rivalry* (Reprint edition, Singapore: Donald Moore, 1965).
Jackson, James, *Chinese in West Borneo Goldfields: A Study in Cultural Geography*, Occasional Papers in Geography, No. 15 (Hull: University of Hull, 1970).
_____ *Sarawak, A Geographical Survey of a Developing State* (London: University of London Press, 1968).
_____ *Planters and Speculators: Chinese and European Agricultural Enterprise in Malaya 1786–1921* (Kuala Lumpur: University of Malaya Press, 1968).
Jacob, Gertrude L., *The Raja of Sarawak: An Account of Sir James Brooke, K.C.B., LL.D., Given Chiefly through Letters and Journals*, 2 vols. (London: Macmillan, 1876).
Jensen, Erik, *The Iban and His Religion* (Oxford: Clarendon, 1974).
Keppel, Henry, *The Expedition to Borneo of H.M.S. Dido for the Suppression of Piracy: With Extracts from the Journal of James Brooke Esq.*, 2 vols. (London: Chapman and Hall, 1846).

―――― *A Sailor's Life under Four Sovereigns*, 3 vols. (London: Macmillan, 1899).

―――― *A Visit to the Indian Archipelago in H.M. Ship Maeander with Portions of the Private Journal of Sir James Brooke, K.C.B.*, 2 vols. (London: Bentley, 1853).

Larkin, John, *The Pampangans, Colonial History in a Philippine Province* (Berkeley: University of California Press, 1972).

Leach, Edmund, *Social Science Research in Sarawak* (London: HMSO, 1950).

LeBar, Frank (ed.), *Ethnic Groups of Insular Southeast Asia* (New Haven: Human Relations Area File Press, 1972).

Lee, Yong Leng, *Population and Settlement in Sarawak* (Singapore: Donald Moore, 1970).

Leigh, Michael B., *The Rising Moon, Political Change in Sarawak* (Sydney: Sydney University Press, 1974).

Lim Teck Ghee, *Peasants and Their Agricultural Economy in Colonial Malaya, 1874–1941* (Kuala Lumpur: Oxford University Press, 1971).

Lo Hsiang Lin, *A Historical Survey of the Lan Fang Presidential System in Western Borneo by Lo Fang-Pai and Other Overseas Chinese* (Hong Kong: Chinese Cultural Institute, 1960).

Lockard, Craig, *From Kampung to City: A Social History of Kuching, Malaysia, 1820–1970*, Monographs in International Studies, Southeast Asia Series, No. 75 (Athens, Ohio: Center for International Studies, Ohio University, 1987).

Longhurst, Henry, *The Borneo Story: The History of the First 100 Years of Trading in the Far East by the Borneo Company Limited* (London: Newman Neame, 1956).

Low, Hugh, *Sarawak; Its Inhabitants and Productions: Being Notes during a Residence in that Country with H.H. The Rajah Brooke* (London: Bentley, 1848; reprinted Singapore: Oxford University Press, 1988).

MacDonald, Malcolm, *Borneo People* (New York: Knopf, 1958).

McCoy, Alfred and de Jesus, Ed. C., *Philippine Social History: Global Trade and Local Transformations* (Quezon City: Ateneo de Manila University Press, 1981).

McDougall, Harriette, *Letters from Sarawak* (Reprint edition, London: Wheldon and Wesley, 1924).

―――― *Sketches of Our Life in Sarawak* (London: Society for Promoting Christian Knowledge, 1882).

McFadyean, Andrew, *The History of Rubber Regulation 1934–1943* (London: George Allen and Unwin, 1944).

Mackie, J. A. C., *The Chinese in Indonesia* (Melbourne: Angus and Robertson, 1974).

Marryat, Frank S., *Borneo and the Indian Archipelago with Drawings of Costume and Scenery* (London: Longman, Brown, Green and Longmans, 1848).

Means, Nathalie, *Malaysian Mosaic: A Story of Fifty Years of Methodism* (Singapore: Methodist Book Room, 1935).

Moor, J. H. (ed.), *Notices of the Indian Archipelago, and Adjacent Countries*, 2nd edition (London: Frank and Cass, 1968).
Morris, H. S., *Report on a Melanau Sago Producing Community in Sarawak* (London: HMSO, 1953).
Morrison, Hedda, *Life in a Longhouse* (Kuching: Borneo Literature Bureau, 1962).
_____ *Sarawak* (Reprint edition, Singapore: Donald Moore, 1965).
Mundy, Rodney, *Narrative of Events in Borneo and Celebes, down to the Occupation of Labuan: From the Journals of James Brooke Esq.*, 2 vols. (London: John Murray, 1848).
Newell, William, *Treacherous River: A Study of Rural Chinese in North Malaya* (Kuala Lumpur: University of Malaya Press, 1962).
Pascoe, C. F., *Two Hundred Years of the S.P.G: An Historical Account of the Society for the Propagation of the Gospel in Foreign Parts, 1701-1900*, 2 vols. (London: Society for the Propagation of the Gospel, 1901).
Payne, Robert, *The White Rajahs* (London: Robert Hale, 1960).
Pelzer, Karl J., *Pioneer Settlement in the Asiatic Tropics: Studies in Land Utilization and Agricultural Colonization in Southeastern Asia* (New York: American Geographical Society, 1948).
Pfeiffer, Ida, *A Lady's Second Journey around the World*, 2 vols. (London: Longman, Brown, Green and Longmans, 1855).
Pringle, Robert, *Rajahs and Rebels* (London: Macmillan, 1970).
Purcell, Victor, *The Chinese in Malaya* (Kuala Lumpur: Oxford University Press, 1967).
_____ *The Chinese in Southeast Asia* (London: Oxford University Press, 1965).
The Ranee Margaret of Sarawak, De Windt Brooke, Margaret Lili Alice, *Good Morning and Good Night* (London: Constable, 1934).
_____ *My Life in Sarawak* (London: Methuen, 1913; reprinted Singapore: Oxford University Press, 1986).
H.H. The Ranee of Sarawak (Sylvia Leonora Brooke), *Sylvia of Sarawak: An Autobiography* (London: Hutchinson, 1936).
_____ *The Three White Rajas* (London: Cassell, 1939).
Rawlins, Joan, *Sarawak 1839-1963* (London: Macmillan, 1965).
Reece, R. H. W., *The Name of Brooke: The End of White Rajah Rule in Sarawak* (Kuala Lumpur: Oxford University Press, 1982).
Richards, A. J. N. (ed.), *The Sea Dayaks and Other Races of Sarawak: Contributions to the Sarawak Gazette between 1888 and 1930* (Kuching: Government Printing Office, 1963).
Roth, Henry Ling, *The Natives of Sarawak and British North Borneo*, 2 vols. (London: Truslove and Hanson, 1896).
Runciman, Steven, *The White Rajahs: A History of Sarawak from 1841 to 1946* (Cambridge: Cambridge University Press, 1960).
Rutter, Owen, *British North Borneo: An Account of Its History, Resources and Native Tribes* (London: Constable, 1922).

BIBLIOGRAPHY 267

—— *The Pagans of North Borneo* (London: Hutchinson, 1929).

—— (ed.), *Rajah Brooke and Baroness Burdett Coutts, Consisting of the Letters from Sir James Brooke, First White Rajah of Sarawak to Miss Angela (afterwards Baroness) Burdett Coutts* (London: Hutchinson, 1935).

Sack, Peter, *Land between Two Laws, Early European Land Acquisitions in New Guinea* (Canberra: Australian National University Press, 1973). .

St. John, Horace, *The Indian Archipelago, Its History and Present State*, 2 vols. (London: Brown, Green and Longmans, 1853).

St. John, Spenser, *Life in the Forests of the Far East*, 2 vols. (London: Smith Elder and Co., 1862; reprinted Singapore: Oxford University Press, 1986).

—— *The Life of Sir James Brooke, Rajah of Sarawak* (London: William Blackwood & Sons, 1879).

—— *Rajah Brooke: The Englishman as Ruler of an Eastern State* (London: Fisher Unwin, 1899).

Sandin, Benedict, *The Sea Dayaks of Borneo before White Rajah Rule* (London: Macmillan, 1967).

Sarawak Shell Berhad, *The Miri Story, the Founding Years of the Malaysian Oil Industry in Sarawak* (Miri: Sarawak Shell Berhad, 1979).

Sharp, Arthur Frederick, *The Wings of the Morning* (London: Greaves, 1954).

Shineberg, Dorothy, *They Came for Sandalwood* (Melbourne: Melbourne University Press, 1967).

Skinner, G. William, *Chinese Society in Thailand, An Analytical History* (Ithaca, New York: Cornell University Press, 1957).

Strauch, Judith, *Chinese Village Politics in the Malaysian State* (Cambridge, Massachusetts: Harvard University Press, 1981).

Tan Giok Lan, *The Chinese of Sukabumi: A Study of Social and Cultural Accommodation*, Modern Indonesia Project (Ithaca, New York: Cornell University Press, 1963).

Tarling, Nicholas, *Britain, the Brookes and Brunei* (Kuala Lumpur: Oxford University Press, 1971).

—— *Piracy and Politics in the Malay World: A Study of British Imperialism in Nineteenth Century Southeast Asia* (Singapore: Donald Moore, 1963).

Templer, John C. (ed.), *The Private Letters of Sir James Brooke, K.C.B., Rajah of Sarawak, Narrating the Events of His Life from 1838 to the Present Time*, 3 vols. (London: Bentley, 1853).

Thompson, H. P., *Into All Lands: The History of the Society for the Propagation of the Gospel in Foreign Parts, 1701–1950* (London: Society for Promoting Christian Knowledge, 1951).

Thompson, Paul, *The Voice of the Past, Oral History* (Oxford: Oxford University Press, 1978).

Tien Ju-Kang, *The Chinese of Sarawak: A Study of Social Structure* (London: London School of Economics and Political Science, 1953).

Trocki, C. A., *Prince of Pirates: The Temenggongs and the Development of Johore and Singapore* (Singapore: Singapore University Press, 1979).

Turnbull, C. M., *A Short History of Malaysia, Singapore and Brunei* (Singapore: Graham Brash, 1980).
Victor, G. and de Bary Nee, Brett, *Longtime Californ', A Documentary Study of an American Chinatown* (New York: Pantheon, 1972).
Walker, H. Wilfred, *Wanderings among South Sea Savages in Borneo and the Philippines* (London: Witherby, 1910).
Wang Gungwu, *A Short History of the Nanyang Chinese* (Singapore: Donald Moore, 1959).
_____ *The Chinese Minority in Southeast Asia* (Singapore: Chopmen, 1978).
_____ (ed.), *Malaysia: A Survey* (London: Pall Mall, 1964).
Ward, A. B., *Rajah's Servant*, Cornell University Southeast Asia Program Data Paper No. 61 (Ithaca, New York: Cornell University Press, 1966).
Ward, A. B. and White, D. C., *Outlines of Sarawak History, 1839–1946* (Kuching: Government Printing Office, 1956).
Warren, James, *The Sulu Zone, 1768–1898: The Dynamics of External Trade, Slavery and Ethnicity in the Transformation of a Southeast Asian Maritime State* (Singapore: Singapore University Press, 1981).
Wertheim, W. F., *East–West Parallels: Sociological Approaches to Modern Asia* (The Hague: W. van Hoeve, 1964).
Wickberg, Edgar, *The Chinese in Philippine Life, 1850–1898* (New Haven: Yale University Press, 1965).
Williams, Lea, *The Future of the Overseas Chinese in Southeast Asia* (New York: McGraw-Hill, 1968).
Willmott, D. E., *The Chinese of Semarang: A Changing Minority Community in Indonesia* (Ithaca, New York: Cornell University Press, 1960).
Willmott, W. E., *The Political Structure of the Chinese Community in Cambodia* (London: The Athlone Press, 1970).
Wu Chun-Hsi, *Dollars, Dependents and Dogma: Overseas Chinese Remittances to Communist China* (Stanford: Hoover Institute, 1967).

Articles

Anonymous, 'The Chinese in Borneo', *The China Review*, Vol. 7 (1878–9), pp. 1–11.
Banks, E., 'The Natives of Sarawak', *JMBRAS*, Vol. 18, No. 2 (August 1940), pp. 49–54.
Blusse, Lenard, 'Chinese Trade to Batavia during the Days of the V.O.C.', *Archipel*, Vol. 8 (1979), pp. 195–213.
Blythe, W. L., 'Historical Sketch of Chinese Labour in Malaya', *JMBRAS*, Vol. 20, No. 1 (1947), pp. 64–114.
Boyd, T. Stirling, 'The Law and Constitution of Sarawak', *Journal of Comparative Legislation and International Law*, Vol. 18, Part 1 (February 1936), pp. 60–70.
Braddell, R., 'Poli in Borneo', *SMJ*, Vol. 5, No. 1 (1949), pp. 5–9.
Butcher, J., 'Towards a History of Malaysian Society, Kuala Lumpur District, 1881–1912', *JSEAS*, Vol. 10, No. 1 (1979), pp. 107–19.

Burns, Robert, 'The Kayans of the Northwest of Borneo', *JIAEA*, Vol. 3 (1849), pp. 140–52.

Chiang Liu, 'Chinese Pioneers, A.D. 1900: The New Foochow Settlement of Sarawak', *SMJ*, Vol. 6, No. 6 (December 1955), pp. 536–48.

Chin, Lucas, 'Archaeological Work in Sarawak', *SMJ*, Vol. 23, No. 44 (1975), pp. 1–7.

―――― 'Impact of Trade Ceramic Objects on Some Aspects of Local Culture', *SMJ*, Vol. 25, No. 46 (July–December 1977), pp. 67–9.

Christie, Jan Wisseman, 'On Po-ni: The Santubong Sites of Sarawak', *SMJ*, Vol. 34, No. 55 (December 1985), pp. 77–87.

Cody, Cecil, 'The Consolidation of the Japanese in Davao', *Comment*, Vol. 7, No. 3 (1958), pp. 26–36.

Coppel, Charles, 'Studying the Chinese Minorities: A Review', *Indonesia*, No. 24 (October 1977), pp. 175–83.

Cruickshank, Bruce, 'Continuity and Change in the Economic and Administrative History of 19th Century Samar', in Alfred McCoy and Ed. C. de Jesus, *Philippine Social History: Global Trade and Local Transformations* (Quezon City: Ateneo de Manila University Press, 1981).

Dalton, G., 'Pepper Growing in Upper Sarawak', *SMJ*, Vol. 1, No. 2 (February 1912), pp. 53–61.

Freedman, Maurice, 'Colonial Law and Chinese Society', *Journal of the Royal Anthropological Institute*, No. 80 (1950), pp. 97–126.

―――― 'Chinese Communities in Southeast Asia: A Review Article', *Pacific Affairs*, No. 33 (1960), pp. 158–68.

―――― 'Immigrants and Associations: Chinese in Nineteenth Century Singapore, *CSSH*, No. 3 (1960), pp. 25–48.

―――― 'Overseas Chinese Associations: A Comment', *CSSH*, No. 3 (1960), pp. 478–80.

Gamba, Charles, 'Chinese Associations in Singapore', *JMBRAS*, Vol. 34, No. 2 (1966), pp. 123–68.

Gardner, Estelle, 'Footnote to Sarawak, 1895–', *SMJ*, Vol. 11, No. 21–2 (July–December 1963), pp. 32–59.

Goldman, Richard, 'The Beginnings of Commercial Development in the Baram and Marudi', *SG*, 31 March 1968, pp. 54–60.

Harper, G. C., 'The Miri Oil Field, 1910–1972', *SMJ*, Vol. 20, No. 40–1 (January–December 1972), pp. 21–30.

Harrisson, Tom, 'Bisaya: Borneo–Philippine Impact of Islam', *SMJ*, Vol. 7, No. 7 (1956), pp. 43–7.

―――― 'Ceramics Penetrating Central Borneo', *SMJ*, Vol. 6, No. 6 (1955), pp. 549–60.

―――― 'Gold and Hindu Influences in West Borneo', *JMBRAS*, Vol. 22, No. 4 (September 1949), pp. 33–110.

―――― 'Interesting Bronzes with Some Ceramics Parallels from Brunei and Sarawak', *SMJ*, Vol. 12, No. 25–6 (1965), pp. 143–7.

―――― 'A Niah Stone-age Burial-jar c-14 Dated', *SMJ*, Vol. 16, No. 32–3 (1968), pp. 64–6.

———— 'Trade Porcelain and Stoneware in Southeast Asia (including Borneo)', *SMJ*, Vol. 10, No. 17–18 (July–December 1963), pp. 222–6.

Harrisson, T. and Jamuh, George, 'Niah: The Oldest Inhabitant Remembers', *SMJ*, Vol. 7, No. 8 (December 1956), pp. 454–63.

Harrisson, T. and O'Connor, Stanley J., 'The Prehistoric Iron Industry in the Sarawak River Delta: Evidence by Association', *SMJ*, Vol. 16, No. 32–3 (1968), pp. 1–54.

———— 'The Tantric Shrine Excavated at Santubong', *SMJ*, Vol. 15, No. 30–1 (1967), pp. 201–22.

Hose, Charles, 'The Metamorphosis of Miri', *British Malaya*, No. 11 (June 1927), pp. 41–6.

Jackson, James, 'Chinese Agricultural Pioneering in Singapore and Johore 1800–1917', *JMBRAS*, Vol. 39, No. 1 (1965), pp. 77–105.

Kaboy, Tuton and Moore, Eine, 'Ceramics and Their Uses among the Coastal Melanau', *SMJ*, Vol. 15, No. 30–1 (1967), pp. 10–29.

Lawrence, A. E., 'Stories of the First Brunei Conquests on the Sarawak Coast', *SMJ*, Vol. 1, No. 1 (February 1911), pp. 121–4.

Lin Wen Tsung, 'The First Ten Years of the New Foochow Colony in Sarawak', *The Sarawak Teacher*, Special History Edition, Vol. 2, No. 2 (1966), pp. 13–14, 18.

Lockard, Craig, 'Charles Brooke and the Foundations of the Modern Chinese Community in Sarawak', *SMJ*, Vol. 19, No. 38–9 (1971), pp. 77–107.

———— 'The 1857 Chinese Rebellion in Sarawak: A Reappraisal', *JSEAS*, Vol. 9, No. 1 (March 1978), pp. 85–98.

———— 'The Early Development of Kuching, 1841–1857', *JMBRAS*, Vol. 49, No. 2 (1976), pp. 107–26.

———— 'Leadership and Power within the Chinese Community of Sarawak: A Historical Survey', *JSEAS*, Vol. 2, No. 2 (September 1971), pp. 195–217.

Logan, J. R., 'Borneo', *JIAEA*, Vol. 2 (1848), pp. 495–7.

———— 'Notices of Chinese Intercourse with Borneo Proper prior to the Establishment of Singapore in 1819', *JIAEA*, Vol. 2 (1848), pp. 611–15.

———— 'Notices of European Intercourse with Borneo Proper prior to the Establishment of Singapore in 1819', *JIAEA*, Vol. 2 (1848), pp. 498–512.

———— 'Traces of the Origin of the Malay Kingdom of Borneo, with Notices of Its Condition when First Discovered by Europeans, and at Later Periods', *JIAEA*, Vol. 2 (1848), pp. 513–27.

Low, Hugh, '*Selesilah* (Book of Descent) of the Rajahs of Brunei', *JSBRAS*, No. 5 (1880), pp. 1–35.

McBain, Audrey, 'Shang and Chou Influence Observed in the Ethnic Art of Borneo', *BMJ*, Vol. 5, No. 1 (1981), pp. 19–32.

Morris, H. S., 'The Coastal Melanau', in Victor King (ed.), *Essays on Borneo Societies*, Hull Monographs on South-East Asia, 7 (Oxford: Oxford University Press for University of Hull, 1978), pp. 37–58.

———— 'How an Old Society was Undermined', *SG*, March 1982.

Moy-Thomas, A. H., 'Economic Development under the Second Rajah (1870–1917)', *SMJ*, Vol. 10, No. 17–18 (July–December 1961), pp. 50–8.
Outram, R., 'The Chinese', in Tom Harrisson (ed.), *The Peoples of Sarawak* (Kuching: Government Printing Office, 1959), pp. 115–28.
Parnell, E., 'The Tributes Paid in Former Days by the then Dependent Provinces of Sarawak', *SMJ*, Vol. 1, No. 1 (February 1911), pp. 125–30.
Pridmore, I., 'Sarawak Currency', *SMJ*, Vol. 7, No. 7 (1956), pp. 111–21.
'Recollections of an Ex-Government Officer, Haji Ahmad Zaidell bin Haji Tahir', *SG*, 31 January 1972, pp. 6–8.
Richards, A. J. N., 'The Descent of Some Saribas Malays', *SMJ*, Vol. 11, No. 21–2 (July–December 1963), pp. 99–107.
'The River Baram. Extracts from a Journal Kept during a Visit to that River in the H. C. Steamer *Pluto*', *JIAEA*, Vol. 5 (1851), pp. 677–90.
'The Saga of Slim', *Salam* (Sarawak Shell Berhad newsletter), 19 February 1955, p. 3.
Sandin, Benedict, 'Descent of Some Saribas Malays and Ibans', *SMJ*, Vol. 11, No. 23–4 (July–December 1964), pp. 512–14.
Saunders, Graham, 'James Brooke and Asian Government', *BMJ*, Vol. 3, No. 1 (1973), pp. 105–17.
Skinner, G. William, 'Change and Persistence in Chinese Culture Overseas: A Comparison of Thailand and Java', *Journal of the South Seas Society*, No. 16 (1960), pp. 86–100.
────── 'Chinese Assimilation and Thai Politics', *JAS*, No. 16 (1957), pp. 237–50.
Scrivenor, John, B., 'Notes on the Geology of Sarawak', *JMBRAS*, Vol. 5, No. 2 (November 1927), pp. 288–94.
Stevens, Peter, 'A History of Kanowit District', Part 2, *SG*, 28 February 1971.
The Siaw Giap, 'The Chinese in Indonesia', *Kabar Seberang*, Vol. 7 (1980), pp. 114–30.
────── 'Group Conflict in a Plural Society', *Revue du Sud-est Asiatique*, No. 1 (1966), pp. 1–31.
Treacher, W. H., 'British Borneo: Sketches of Brunei, Sarawak, Labuan and North Borneo', *JSBRAS*, No. 20 (1889), pp. 13–24; No. 21 (June 1890), pp. 19–121.
Trocki, Carl, 'The Origin of the *Kanchu* System, 1740–1860', *JMBRAS*, Vol. 49, No. 2 (1976), pp. 132–56.
Wang Gungwu, 'The Nanhai Trade, A Study of the Early History of Chinese Trade in the South China Sea', *JMBRAS*, Vol. 31, No. 2 (June 1958), pp. 1–135.
Wang Tai Peng, 'The Word Kongsi: A Note', *JMBRAS*, Vol. 3, No. 1 (1979), pp. 102–5.
Ward, Barbara, 'A Hakka *Kongsi* in Borneo', *Journal of Oriental Studies*, Vol. 1, No. 2 (July 1954), pp. 358–70.
Warren, James, 'Joseph Conrad's Fiction as Southeast Asian History:

Trade and Politics in East Borneo in the Late 19th Century', *BMJ*, Vol. 4, No. 1 (1977), pp. 21–34.

Wickberg, Edgar, 'The Chinese Mestizo in Philippine History', *Journal of Southeast Asian History*, No. 5 (1964), pp. 62–100.

―――― 'Early Chinese Economic Influence in the Philippines, 1850–1898', *Pacific Affairs*, Vol. 37, No. 3 (1962), pp. 275–85.

Willmott, W. E., 'Congregations and Associations: The Political Structure of the Chinese Community in Phnom Penh', *CSSH*, Vol. 2 (1969), pp. 282–301.

Wong Lin Ken, 'The Trade of Singapore, 1819–69', *JMBRAS*, Vol. 33, No. 4 (December 1960), pp. 5–315.

Yong Ching Fatt, 'A Preliminary Study of Chinese Leadership in Singapore, 1900–1941', *Journal of Southeast Asian History*, Vol. 9, No. 2 (1968), pp. 258–85.

Young, G. V. C., 'The Transitional Law Period in Sarawak, 1927–1941', *SG*, 30 November 1964, p. 283.

Chinese-language Sources

Chen Khoon Yan, 'A Brief History of the Opening of Spaoh', in Teochew Association, Kuching, *Centenary Volume* (Kuching: Teochew Association, 1965).

Foochow Association, Sibu, *The Fiftieth Anniversary of the Foochow Settlement in Sibu, 1901–1950* (Hong Kong: Chiao Kuong Press, 1951).

―――― *The Sixtieth Anniversary of the Foochow Settlement in Sibu, 1901–1960* (Hong Kong: Chiao Kuong Press, 1961).

Kho Chong Soo, *The British Borneo Year Book, I* (Kuching: 1952).

Kuching Hokkien Association, *Centenary Souvenir Magazine* (Kuching: Borneo Publication, 1980).

Lau Tzy Cheng, *Borneo History* (Sibu: Rejang Book Company, 1964).

―――― 'Chinese Publishing Activities in Sarawak', *Journal of South Seas Society*, Vol. 20, No. 1–2 (1965), pp. 1–12.

―――― *Fung Hsia Miscellaneous Articles* (Singapore: Archipelago Cultural Publisher, 1980).

―――― *Wong Nai Siong and the New Foochow* (Singapore: Nanyang Institute, 1979).

Liew Peck Kwee, *A History of the Hoppo Chinese with Special Reference to Sarawak* (Singapore: Tung Yet Publishing House, 1978).

―――― 'The Relationship between the Bau Kongsi and the Brooke Government', in Teochew Association, Kuching, *Centenary Volume* (Kuching: Teochew Association, 1965).

Ong Ngee Guan, 'Teochew Chinese in Simanggang', in Teochew Association, Kuching, *Centenary Volume* (Kuching: Teochew Association, 1965).

Teochew Association, Kuching, *Centenary Volume* (Kuching: Teochew Association, 1965).

Bibliographies

Cotter, Conrad P., *Bibliography of English Language Sources on Human Ecology, Eastern Malaysia and Brunei*, 2 vols. (Honolulu: Asian Studies Department, University of Hawaii, 1965).

―――― 'A Guide to the Sarawak Gazette, 1870–1965', Ph.D. thesis, Cornell University, 1966.

Harrisson, Tom, 'Historical and Related Sources for Sarawak', in K. G. Tregonning (ed.), *Malaysian Historical Sources* (Singapore: University of Singapore Press, 1962), pp. 105–12.

Leigh, Michael B., *Checklist of Holdings on Borneo in the Cornell University Libraries*, Cornell University Southeast Asia Program Data Paper No. 62 (Ithaca, New York: Cornell University Press, 1966).

Nevadomsky, Joseph John, and Li, Alice, *The Chinese in Southeast Asia: A Selected and Annotated Bibliography of Publications in Western Languages*, University of California Occasional Papers of the Centre for South and Southeast Asian Studies, 1970.

Pearson, James Douglas, Wainwright, M., and Matthews, M. A., *A Guide to Manuscripts and Documents in the British Isles Relating to South and Southeast Asia* (London: Oxford University Press, 1965).

Index

ABAN JAU, 113; *see also* Tinjar River
Aban Nipa, 74–5
Absconding from debts, 119
Acculturation, of Chinese traders, 134–5, 223
Adeh, in Marudi, 75, 105
Advances, of goods, 91, 115
Ah Sam (Assam), Chinese boat hawker, 135–6
Ah Toon, 'living among Dayaks', 123
Anglo-Saxon Petroleum Company: 193–4; obtained coolies from Singapore, 193
Antimony, 23
Autonomy: of Batang Krian traders, 68; bazaar, 129; of *ulu* traders, 138

BALLEH RIVER, 71–2
Banjermassin, 54
Baram District census, 75–6
Baram River, 72–4
Baram River, trade prospects, 74
Baring-Gould: comments on Foochows abandoning padi and pepper cultivation for rubber, 155–6; comments on Land Office, 164
Barter exchange, 93, 108; *see also* Tay Bak Choon
Batang Bulungan: Kayans coming to trade, 112
Batang Kayan: Kayans coming to trade, 112
Batang Lupar: 65, 66, 67; pioneers, 79; *see also* pioneering life
Bau: 22, 24, 37, 40, 42; Brooke presence in, 41; collapse as autonomous Hakka mining district, 38, mining and agricultural settlements, 24; *see also* Chinese in Bau; Hakkas; *Kongsi*
Bejalai, 109
Belaga, 72, 112, 126
Betong, 67–8

Bezoar stones, 102
Bidayuhs, 13; *see also* Land Dayaks
Binatang: Foochows in, 160–1; *see also* Foochows
Bird's nest, 102
Blood-brother pact, 95–6, 99 n.
Boat hawkers: 77, 78–9, 80, 126; free from official interference, 126; as marginal occupation, 90; dangers faced by, 80, 81, 82, 86, 95–6; renewal of monthly licences, 125
Boating mishaps, 80–1; *see also* Boat hawkers
Borneo, 3, 4, 50, 52, 54, 55, 58, 100, 102
Borneo Company, 33, 37, 38, 39, 40, 48 n., 181–2
Bowring Treaty, 1855, (Siam), 222
Brooke, Charles: 15, 38; agreement with Wong Nai Siong, 142; attitude to Chinese, 157 n.; attitude to Sadong coal miners, 188; and Chinese Protectorate, Singapore, 183–4; on salt control, 126; and secret societies, 41–2; and the summoning of native debtors, 116; on the use of rattan whip on Chinese labourers, 188; visit to Bau, 46; visit to Marudi, 75; visit to Simanggang, 137
Brooke, Charles Vyner: 15; attitude to Chinese, 162
Brooke, James: acquisition of Sarawak, 15, 29; plans for Sarawak, 203; and sovereignty over the Chinese, 207; view on Bau *kongsi*, 24; view on justice, 208
Brooke administration as protector of native interests, 148
Brooke attitudes: to Chinese immigration, 173; to Chinese traders, 122, 124; to commercial development, 181; to Foochows, 169;

INDEX 275

to intermarriage, 137; to gambling and drinking, 85
Brooke economic philosophy, 191; *see also* Brooke attitude to commercial development
Brooke policies, 76
Brooketon (Muara) coal mines, 192
Brunei, 15, 53, 54, 55, 56
Brunei, Sultan of, 63, 74
Brunei Malays, 13, 74, 75, 95, 108, 118, 133

CANNONS, 108
Cantonese: 10, 119 n.; engaged in timber logging in Sarikei, 119 n.; in Sadong colliery, 189; settlement in Sibu, 177 n.; *see also* Chinese labourers of Sadong colliery
Ceramics, 52; *see also* Jars
Cession of Sarawak, 15
Chain migration of Tay clan to Simanggang, 79–80
Chan Ah Koh, 39
Chan Wey Yuen, 196
Chao-anns, 10
Chau Ju-Kua, 53
Chau Soon, 'living among Dayaks', 123
Cheyne Ah Fook, Chinese court writer, 206
Chin Ann and Company, 45
Chinese, vii, viii, 1, 3, 5, 8–10, 24, 26, 28, 37, 38, 46, 50, 52, 53, 59, 60, 69, 74, 75, 76, 82, 85, 86, 87, 91, 102, 104, 109, 116, 118, 122, 128, 130, 132, 133, 134, 168, 181, 188, 203, 219, 220, 222, 223, 227, 228, 229, 230
Chinese, repatriation of jobless, 172
Chinese accommodation of native clients, 134; *see also* Acculturation, of Chinese traders
Chinese attitude to business, 132–3
Chinese collective experiences, 230
Chinese coolies at Miri oilfields: 194; deaths due to malaria and beriberi, 195; coolie lines, 196; on strike, 197–9
Chinese Court, 204, 210, 256–7
Chinese customary law, 210–11; *see also* Chinese Court; Chinese family law

Chinese economic contributions, 203, 211, 215, 216, 228
Chinese education in the Lower Rejang, 151–2; *see also* Chinese schools
Chinese ethnic identity, 85, 153
Chinese family law, 210; *see also* Chinese customary law
Chinese in Bau: arrival, 6; Chinese census in Bau in 1870, 39; Chinese–Land Dayak relations, 28, 29; *see also* Bau; Hakkas; *Kongsi*
Chinese in rural areas, x n., 228
Chinese labourers at Sadong colliery: 181–5; accidents, 186; absconding, 186–7; coolie lines, 184–5; difficulty in obtaining labour, 184; labour force in 1891, 184; piece-rate system 188–9; recruited through Chinese Protectorate in Singapore, 183; *Sarawak Gazette* medical reports on working and living conditions, 185–6; system of procuring indentured labour, 187; coolies on strike, 188–9; *see also* Cantonese, Hakkas
Chinese patronage of opium, gambling, pawn and arrack farms, 212
Chinese Protectorate, Singapore, 183–4, 189; *see also* Singapore
Chinese Rebellion of 1857, 47 n.; *see also* Kongsi–Brooke war
Chinese schools, 85, 87; *see also* Chinese education in the Lower Rejang
Chinese shopkeepers, 114, 117; *see also* Chinese traders
Chinese temple, 86, *see also* Kwan Yin; Tua Pek Kong
Chinese traders: 12, 55, 82, 90, 107, 110, 111, 112, 130, 132, 133, 137, 153, 220, 223; on the Baram River, 76; in bazaars, 107; competition among Chinese traders in the Rejang, 111; in Long Akah, 125; in Lubok Antu, 82; in Lio Matu, 125; in Marudi, 105; in the Rejang, 110; on Sungei Tinjar, 125; wanting to live among the natives, 124
Chinese–Iban intermarriage, 135, 136,

137–8; *see also* Chinese–Land Dayak intermarriage
Chinese–indigenous relations, 228; *see also* Chinese accommodation of native clients; Acculturation of Chinese traders
Chinese–Land Dayak intermarriage, 28; *see also* Chinese–Iban intermarriage
Chinese–Malay co-operation in Belaga, 134
Chinese–Malay trading rivalry, 133; competition with Malay boat hawkers, 116
Cholera epidemics, 102
Chop Chiap Heng, 128
Coal, 181–2; *see also* Chinese labourers on Sadong colliery
Commission of Inquiry, into James Brooke's conduct, 1854, 32
Copper coins, 108
Court, Bankruptcy, 209
Court, General, 208–9
Court, Police, 208–9
Court, Supreme, 208
Court cases, 210
Court of Debtors, 209
Court of Requests, 209
Credit system: 114; in the Baram, 117–19; *see also* Chinese traders
Crespigny Claude Champion de, 74
Cultivation, shifting, 163, 230

DATCHING, FRADULENT, 127; *see also* Fraudulent weights and measures
Davao, Japanese settlers in, 224
Dayak deposits, 91, 114, 116
Dayaks, definition of, 11
Depression of 1929, 172
Dialect group fights, 131–2, 177 n., 189–90
Digby, K. H., comment on system of procuring indentured labour, 187
Division, definition of, xiv, 4
Duties, import and export, 214–16

ECONOMIC TRANSFORMATION, OF RIVERINE ECONOMIES, 107
Economy, self-sustaining, 212
Engkabang: 102; engkabang boom in

Batang Lupar, 102–3
Engkilili, 67
Ensurai, 145
Ethnographic background, 6–15
Europeans: 14–15; in Miri, 197, 202 n.; European staff in Anglo-Saxon Petroleum Company, 193

FARMS, 211–14; *see also* Monopolies
Fire, in Sibu, 247–8
Foochows: 9, 223, 225; arrival in 1901–2, 143, 145; associations, 176–7; attitude to cultivation, 163; debate on migrating to the Rejang, 144; expansion into sectors of the modern economy, 229; faced isolation and rivalry, and encountered tension, 174; first group of Foochows, 143; outflow of Foochows as a result of slump of 1920–1, 170; planting padi, 146–7; prevalence of illness and death, 146; switch to pepper and rubber cultivation, 149; *see also* Chinese education in the Lower Rejang; Methodist Church; Sibu
Forts: 63; Fort Alice, 65–6; Fort Hose, 74; Fort James, 65; Nanga Balleh Fort, 71
Fraudulent products, 128
Fraudulent weights and measures, 127–8; *see also* Datching
Freeman, Derek, 163
Frontier, extension of, 24, 63, 215; riverine, 69

GAMBIER, 46
Gambling: 84, 85, 189, 196; gambling dens, 85
Goodwill between Chinese trader and native client, 83, 98–9 n., 134
Government assistance for unemployed Foochows, 170, 172
'Guarantee system' for new arrivals, 168
Gun-selling prohibition, 126–7
Gutta percha, 100, 109

HAKKAS: 8; in Bau, 27, 31, 33, 37, 39, 40, 41, 43, 219–20; and Borneo

Company, 43, 45–7; and cash cropping, 47; and Cantonese clash in Sadong colliery, 189; in Kalimantan, 18–20; and Land Dayaks clash, 46; *see also* Bau; *Kongsi*
Helms, Ludvig, 25
Henghuas, 10, 170
Hokkiens: 9, 220–1; competing against Teochiu traders, 130; extensive trading network stretching to Kuching, 132; as pioneering traders, 78; as traders on the Rejang River, 86; in Kanowit, 69; in Kuching, 25, 31, 37, 38, 204; in Marudi, 74
Hong Kong, 119 n., 195
Hoover, James: 143, 150, 151–2, 162, 167, 175–6; description of land rush in Binatang, 167
Hose, Charles, 192–3; see also Credit system in the Baram

IBANS: 12–13; attack on Foochows, 168; attitude to Foochows, 162; attitude to rubber planting, 164; attitude to swidden farming, 163–4; Batang Ai Ibans, 109; and Chinese boat hawkers, 79; contribution to revenue, 215; dependence of Batang Lupar Ibans on Chinese traders, 110; and head-hunting, 82; land *adat*, 167–8, 178 n.; Lubok Antu Ibans, 82; migration to Lower and Middle Rejang, 163; pioneering Ibans on Rejang frontier, 145; Iban rangers, 131; reliance on Ibans for procurement of jungle products, 102; selling fraudulent products, 128; and superstitions, 83; and trade, 84; warned against allowing Chinese traders to live in longhouses, 123, 138 n.
Indian traders, 209
Indians, 52
Interception of native products, 115

JARS, 52; *see also* Ceramics
Junks, Chinese, 55, 80–1, 106–7, 109–10, 244–5

KABONG BAZAAR, 244

Kalimantan, 18, 22
Kanchu, 205, 206–7, 217 n.
Kanowit, 69, 71
Kapit, 69, 72
Kapitan, 204, 205–7
Kapitan China General, 204
Kayans, 13–14, 110, 111–12, 113, 117–19
Kelabits, 14, 113
Kenyahs, 14, 113
Kong Aik Bank, collapse in Singapore and repercussions, 114
Kongsi: 18, 20; autonomy, 30, 31; *kongsi*-Brooke rivalry, 29–31; *kongsi*-Brooke war, 37–8; economic organization, 27; 'Fifteen Company', 32, 34 n.; Ho-shun *kongsi*, 20, 21; *hui*, 20; *kongsi*-house, 26, 34 n.; Lan-fang *kongsi*, 21; mining technology, 27; post-1857 *kongsi*, 39–40; post-1857 *kongsi* and organizational structure, 40–1; San-tiao-kou *kongsi*, 21; San-tiao-kou *kongsi* refugees in Bau, 22, 23, 24; Shak Luk Mun *kongsi*, 40; *shan-sha*, 19; self-government, 25–6; 'Twelve Company', 23, 25, 31, 219–20; *see also* Bau; Hakkas
Krian River, 65
Kuala Baram bar, 106
Kuching: 5, 38–9, 43, 49 n.; control over *ulu* trade, 92; *see also* Hokkien and Teochiu traders in Kuching
Kuching General Hospital and treatment of Sadong colliery workers, 185
Kutien Association, 176–7
Kwan Yin, 86
Kwang Hua School Alumni Association, Sibu, 177
Kwang Yung Serh Association, Sibu, 177

LAND: attitudes to usage: 230; Foochow attitude, 164; Iban attitude, 164
Land concession: Lower Rejang, 153; Binatang, 166
Land Dayak land rights, 45–6
Land disputes, 161–8, 224

INDEX

Land friction: in Binatang, 167–8; in Lundu, 45
Land Office, 165–6; responsibility placed on Land Office because of land competition and disputes, 165–6
Land Orders, 1931, 169
Land Ordinance, 1948, 169–70; *see also* Law of Sarawak Order, 1928
Land Settlement Order, 1933, 169; *see also* Law of Sarawak Order, 1928; Land Ordinance, 1948
Land surveying, 164
Land titles: in Lower Rejang, 154; in Binatang, 167
Landscape, physical, 3–6
Lau Ka Too, 217
Law Kian Huat, 204
Law of Sarawak Order, 1928, 210; *see also* Land Settlement Order, 1933; Land Ordinance, 1948
Lee Kai Ku, 44
Lee Sui, 44
Lee Yok Seng, 197
Legislation, on broken intermarriages and custody of children, 137
Lek Chiong, 143
Letter Books, 1
Liew Shan Pang, 23
Lim Goon Teng, 89
Lim Gu Piao, 84
Lin Wen Tsung, 145–6
Ling Kai Cheng, 151
Lingga, 66
'Living among Dayaks', 122–5; regulations in the Baram, 125; *see also* Chau Soon; Ah Toon
Lo Pang Po, 21
Lockard, Craig, vii, ix n., 225
Long Akah, 74
Long Lama, 76
Low, Hugh, 57–8
Lubok Antu, 66–7; trade in, 82
Lundu, 35 n.; *see also* Land friction in Lundu

MACKIE, J. A. C., 224–5
Malays: 11–12; in the Baram River, 57; boat hawkers, 116; commerce, 222; ethnic origins, 59–60; traders, 55–7
Marudi bazaar, 75

Melanaus, 17, 215
Mestizo, 223
Methodist Church: 149–51; expansion, 174–6; lay preachers, 175; schools, 151–2, 223, 225; *see also* Hoover, James
Miri, 5, 193, 194, 196, 197, 202 n.
Miri Recreation Club, 197
Monopolies, 212; *see also* Farms
Mundy, Rodney, 26
Murder cases of boat hawkers, 81, 86–7, 89–90; *see also* Boat hawkers
Muslim–animist relationships, 6

NIAH, 75, 76
Niah Caves, 50
Neo Teck Hua, 88–9

'OCCUPATION TICKETS', 156, 165, 178 n.
Oil, 192–3; *see also* Chinese coolies at Miri oilfields
Oil drilling as occupational hazard, 195; *see also* Chinese coolies at Miri oilfields
Ong Ewe Hai, 204
Ong Tiang Swee, 204
Opium: 28; tax, 29, 32; smoking, 158 n., 189; *see also* Farms; Monopolies
Oral history, 2–3, 77, 96, 143, 146, 195
Orders issued, which concerned the Chinese, 208
Overseas Chinese, viii

PARTI BANGSA DAYAK SARAWAK, 229
Penans, 14, 102
Pengiran Makota, 15
Pengiran Muda Hashim, 15, 29
Pepper, 43, 44, 45
Peranakan, 223
Pigafetta, 54
Pioneering conditions: on Bau frontier, 25, 28, 41; in Lower Rejang, 146
Pioneering initiative of rural traders, 221
Pioneering life: in the Baram, 95; in the Rejang, 89; of the Chinese traders, 77–9
Pioneering spirit in building bazaars, 67

P'o-ni, 53–4
Pringle, Robert, ix n.
Probate Books, 89, 95
Probate Cases: Ah Hiong, 93; Ah Wa, 94; Ah Wai, 94; Bee Ah Wai, 95; Chap Long, 91; Chin Poh, 92; Lee Ah King, 95; Oh Kee, 95; Vi Ah Cham, 93, *see also* Probate Books
Pui Shin Wen, 43–4
Pulang Gana, 12

RAJAH'S ORDER BOOKS, 208
Rapids: impediment to trade and transport, 88, 89, 96
Rattan, 100–1; trade in the Rejang, 103–5; in Marudi, 105
Reciprocity, 57, 98–9; *see also* Goodwill between Chinese trader and native client
Rejang River, 69
Rejang Settlement Order, 249
Residents: conduct, 209–10; hearing court cases, 210
Rice cultivation in the Rejang, 253–5
Rice shortages, 110; in the Rejang, 156–7
Rivers, 5–6, 63; *see also* Baram River; Batang Lupar; Rejang River; Rapids
Rubber: boom, 169; high prices, 153; fluctuating prices, 153; in the Lower Rejang, 149; low price in 1920, 156; price collapse in 1929, 172; rubber restriction, 1934, 173–4; slump in 1920 and 1921, 170; tapping routine, 155

SADONG COLLIERY, 181, 183–4, 191–2; *see also* Chinese labourers on Sadong colliery
Sago, 13, 215
Salt sales control of, 126
Santubong, 52, 61 n.
Saratok, 68, 244
Sarawak Gazette: accounts, 1, 77, 89, 106; reports on pepper plantations, 39, 45; report on agricultural conditions in the Lower Rejang, 146; report on Sadong coal mine, 182; report on Sadong coal mine labourers, 185; report on Miri oilfield labourers, 195; report on Miri oilfield strikes, 199
Sarawak Government Gazette, 1
Sarawak National Party, 229
Sarawak nationalism, 229
Sarawak Opium Farm Syndicate, 212–14; *see also* Farms; Monopolies
Sarawak River, 50, 52, 58; *see also* Santubong
Sarawak United People's Party, 229
Saribas River: 65, 67; pre-Brooke trading arrangements, 65
Sarikei: Foochows in, 161; *see also* Foochows
Secret societies, 41–2, 49 n., 190, 198
Secretariat for Chinese Affairs, 204–5
Seng Joo, 79–80
Senior Asiatic Boat Club, 197
Shipping difficulties in the Baram, 106
Shophouse, 16 n.
Shophouse inspection, 129
Shophouse structure, 66
Shophouse taxes, 129
Siaw Kee Siong, 87–8
Sibu, 5, 69, 71, 143, 145, 166, 170, 171–2
Sibu Chinese Chambers of Commerce, 176
Sibu Foochow Commercial Recreation Club, 177
Sibu Methodist Philanthropic Association, 177
Simanggang, 5, 65–7
Simanggang bazaar, 66
Simunjan, 5, 184, 192
Sing Chio Ang Church, 150; *see also* Methodist Church
Singapore: *sinkheh* coolies from, 46–7, 193; trade with Borneo, 58, 60
Siniawan, 24
Sinkheh, 78, 80, 146, 186–7, 192
Skinner, G. William, 225
Skrang River Ibans, 82
Sleh, Haji, 59–60
Song, 72
Sovereignty, 29; *see also* Brooke, James, and sovereignty over the Chinese
Spaoh bazaar, 68
St. John, Spenser, 24–5, 108

Strike, 197–9; *see also* Chinese coolies on Miri oilfields
Sulu, 55, 222
Sungei Merah (Seduan), 145
Syn Chin Joo, 126, 134

TAI PEK KONG, 26; *see also* Tua Pek Kong
Tanjong Lutong submarine oil pipeline, 194; *see also* Chinese coolies at Miri oilfields
Tay Bak Choon, 82–3
Tay Ing Boon, 83–4
Tay Sze Heng, 79
Teo Chong Loh, 86
Teochius: 10, 43; as pioneering traders, 78; in the Rejang River, 86; in Sibu, 172; in Simanggang and Lubok Antu, 66–7; at Skrang River mouth, 65; in trade, 220–1; traders in Kuching, 25, 31, 37, 38, 203; trading network stretching to Kuching, 132
The Siaw Giap, 224
Tien Chu Kim, 95–6
Tien Ju-Kang, vii

Timber-logging disputes, 230
Ting Kwong Dou, 143
Tinjar River, 113; *see also* Aban Jau
Totok, 226; *see also Peranakan*
Trade: 220–1; as Hokkien and Teochiu pioneering activity, 55; and raiding, 56, 57; and politics, 57; 'forced', 56–7
Transvaal, Chinese labourers in, 227
Tributary relations, 50
Tua Pek Kong, 75, 189; *see also* Tai Pek Kong
Tuai rumah: permission to trade, 125–6
'*Tukang ayer*', 195–6
Tutoh River, 75

UNDANG-UNDANG, 207
Undup River, 123

WAGE LABOURING, 226
Wang Tai Peng, 26, 220
Wayang shows, 85, 190
Wong, James, Datuk Amar, 231 n.
Wong Gin Huo, 206
Wong Nai Siong, 142–3, 147, 149; *see also* Foochows